WHAT

IS A

WOMAN?

WHAT

IS A

WOMAN?

ONE MAN'S JOURNEY
TO ANSWER THE QUESTION
OF A GENERATION

MATT WALSH

What Is A Woman?

Copyright © 2022 by Matt Walsh

ISBN: 978-1-956007-00-8

Jacket Design by David Fassett

First Edition

Published by DW Books
DW Books is a division of Daily Wire

Daily Wire
1831 12th Avenue South
Suite 460
Nashville, TN 37203
www.dailywire.com

PRINTED IN THE USA

CONTENTS

A SIMPLE QUESTION

I WAS STANDING in front of a naked man.

This wasn't exactly how I thought my career would go. But I was determined to leave no stone unturned.

"Being a sixty-six-year-old man," he said, "I mean, there's the obvious things, you know, that physically make a woman."

I was in San Francisco's Castro District, probably the queerest place in America. Rainbow flags were everywhere. Luckily, our little corner of the street was free of human feces and needles. And I guess technically the man I was talking to wasn't completely naked. He *was* wearing a sock. Just one.

"There are features of a woman that I'm attracted to and that..." he paused for a moment. "I'm probably fumbling over the answer."

He wasn't the only one fumbling for an answer.

I had gathered up a documentary crew and had been asking everyone I could find what I thought was an absurdly—almost insultingly—simple question: What is a woman? Yet out of confusion or fear or sometimes even anger, almost nobody could answer the question.

I went to Times Square in New York, hoping maybe somebody on the other side of the country could help me out.

"What, specifically, makes someone a woman?" I asked a guy on the street.

"I wouldn't know what to answer that," he said.

Ok then. I turned around and saw a couple walking by. Maybe I'd have better luck with two people instead of one, so I asked them what their answer would be.

"Ummm... I... that is a hard question," the girl said.

"*Is it* a hard question?" I asked.

"Yes," she said with a drawn-out pause.

"Why is it hard?" I followed up.

"Because I personally believe gender is fluid. I think there is a distinguishment between sex and gender," she responded. "So, I don't know if there is a picture-perfect way to describe a woman."

That does beg the question what women's rights, women's issues, women's products, women's clothing, women's literature, women's studies, or the women's march actually means. I soldiered on.

"What is a woman?" I asked a woman walking down the street in Hollywood, California.

"Ummm..." she began before a long pause. "A choice. Ok, not like a choice but like a, uh, option. Like, you're... I suppose it's because you're not determined from the moment you're born. You're freer."

Free from what, I wondered. And I still didn't have an answer. I was compelled to try again.

"So, what is a woman?" I asked a woman in Nashville, Tennessee.

"A woman is someone that likes to be pretty and think of themselves as a delicate creature," she said.

"I'm pretty and delicate," I said. "I could be a woman too?"

"Yes, you could. If you wanted to be, you could," she answered.

I turned to another woman. Well, I guess I wasn't sure. She looked like a woman and spoke like a woman, but I didn't ask her if she thought of herself as pretty and delicate. "How would you define the word 'woman'?" I asked.

"I think someone who identifies as a woman," the... person... answered.

"Someone who identifies as a woman is a woman?" I asked.

"Yeah."

"Alright, that's it?"

"That's it!"

Identity. Being delicate. An option. There seemed to be a million answers for what is a woman. Oddly enough, none of them featured biology or DNA or reproduction. Some people I asked rejected the idea that how you were born has any claim over your sex.

And some asserted that anyone who is not a woman (whatever that word means) doesn't even have a right to answer the question.

"Why are you asking a gay man as to what it means to be a woman?" someone said to me accusingly as I had just finished up talking with a rather flamboyant gay man. The person approaching looked to be a biological male trying to pass as a woman—a so-called trans woman, as I learned. He had thick features, sported a purple fedora, and wore a black shirt with the words "Eat. Halloween. Repeat." in blocky letters. "You should be asking women what it means to be a woman, especially trans women," the man continued.

"I'm asking all kinds of people," I responded guilelessly. "Can anyone have an opinion about it?"

"Only people who are a woman," he said. "Gay men don't know nothing about what it means to be a woman."

"So gay men shouldn't have opinions on what a woman is?" I asked. "Have you told gay men here in San Francisco that they're not allowed to talk about this?"

"No, but it's not like I go around talking about what a gay man is allowed to be," he answered.

"I didn't say what you're allowed to *be*," I clarified. "I'm asking people their opinions on what a woman is."

"Same difference," he said. "If somebody was to come and ask me what a gay man is, I'm like, 'Go ask a gay man.'"

Hold on, I thought—does identity determine who is allowed to speak? I asked him: "So, you're saying if you're not a woman, then you shouldn't have an opinion?"

"Where does a guy get a right to say what a woman is? Women only know what women are," he said.

"Are you a cat?" I asked.

"No."

"Can you tell me what a cat is?"

He looked down, then glanced at my cameraman through his sunglasses, holding back a grimace. The walls of the intellectual corner he had painted around himself were closing in fast.

"You know what, this is actually a genuine mistake, and I am sorry I even came up here." He walked away.

"Do you want to tell us what a woman is?" I called after him. He made no response.

AN EPIC JOURNEY

What is a woman? For the past eight months, I have devoted nearly all of my waking hours to see who can answer this simple question. Honestly, it's odd to even feel the need to ask.

For most of my life, I thought everyone knew the answer.

Maybe you did too. It's not like we had to pull out Webster's Dictionary to figure it out. Women are adult human females. They have XX chromosomes. They can bear children and give birth. They're not necessarily nicer than men, but they sure are better looking. Even if you didn't know the science or use the exact right words, you could point a woman out pretty easily. By nature, they look and act differently from men.

But as with so many things in life, I found in my journey that the educated set of our civilization has developed an entire ideology confounding this simple question—and that confusion has percolated throughout society to the point that a dwindling few people can tell up from down anymore.

It's not like relativist gender theorists are aware of a large number of nuances or qualifications that old-fashioned rubes like us can't understand. The honest truth is that, in all their learning, they unlearned common sense.

In eerie uniformity, the world's most credentialed experts and powerful public personalities have started telling us that men can become women and women can become men or even that people can become something in between a man or a woman. Or beyond? Outside of? Alternative to? They don't seem terribly clear when you push them beyond the normal talking points.

We all thought being a woman had something to do with biology, but the nation's top experts keep assuring us that is definitely not the case. Maybe it's when people have certain feminine qualities. But aren't gender roles a social construct? Perhaps like that one woman on the street told me: it's when anybody thinks he or she is a woman, then that person becomes a woman. But thinking I'm Superman doesn't make me fly. Is there a difference between being a woman because you think you're a woman and being Superman because you think your Superman? Is it sexist to make such a comparison without assuring everyone that all women are superheroes? Is Superman actually a woman?

I had asked all sorts of people out in public and hadn't gotten any clarity, so I decided to turn to the experts. Trouble is, the more I read from the so-called experts, the more confused I became. All of a sudden an entirely new glossary that I had never been taught in my English classes was being tossed about with such confidence (and condemnation of those who couldn't keep up) that my head started to spin. Words and phrases bubbled up everywhere such as gender dysphoria, gender reassignment, preferred pronouns, assigned sex, metoidioplasty, she/her, he/him, they/them, em/eirs, ze/hir.... To avoid the confusion, people everywhere started abusing the third-person plural pronoun "they" like it was one of Dr. Fauci's test puppies.

Anyone with sanity and a spine can see pretty quickly that gender theory is a load of bull. But the ideology has taken hold so thoroughly and so quickly that it's easy to assume there must be something to it. That assumption, it turns out, is wrong.

Sure, in our age of postmodern reinvention where true freedom is thought to be the ability to "define one's own concept of existence, of meaning, of the universe, and of the mystery of human life," gender theory has a certain appeal.[1] The ultimate act of self-definition is to declare by feeling and fiat that you are male even though every cell in your body is coded "female." That, and being trans, is a very quick way to become the left's coolest, latest victim class. Confused young people uncertain about their bodies (something we used to just call puberty) can now say they are transgender. Once proclaimed, they immediately feel the warm, affirming embrace of innumerable anonymous online personalities eager to aid them in their transition, coupled with the satisfaction of knowing that, unlike all those spiteful, privileged people, they really are oppressed.

However, despite its popularity, the ideology is astoundingly thin. It's like pop music or blockbuster movies. It drives sales, but there's not really much meaning in it all. Proponents may use a complicated vocabulary, boast uniform elite acceptance, and assert that they have the blessings of almighty science. But their words have no substance.

That brings me to my ultimate question. It's a simple question, but one that the gender ideologues cannot answer. I know, because I asked them.

For months I sought out experts in every field in the gender theory world—doctors, therapists, psychiatrists, psychologists, politicians, activists, and transgender people themselves—and I put the question to them directly. I didn't waltz in with an agenda. In fact, for the sake of inquiry, I was willing to take exactly what they said at face value. I wanted their answers to speak for themselves. I wasn't trying to embarrass, entrap, or lead on any of these people. If something they said didn't make sense right away, I presumed it was because I didn't understand.

But as time went on and as the interview hours ticked up, I began to realize that the problem wasn't my understanding. Nothing they said made any sense. They contradicted themselves, sometimes within minutes of speaking. Their terms were fuzzy and quickly discarded. They made bold proclamations but could never say exactly why their proclamation was so. Their arguments collapsed, not with hammer blows or a hard push, but with soft questions, because the core of gender ideology is hollow. Confusing words simply mask the fact that there is no "there"

there. But I don't want you to take my word for it. You should hear it from the experts themselves.

Are you still wondering what is a woman? Well, join me, and I'll let the best and the brightest answer the question themselves.

THE HISTORY OF GENDER THEORY

TRACING THE ORIGINS of gender theory was no easy feat. After all, the male-female difference seems baked into human existence. Just about every culture out there, from the Greeks to tribes in Africa to the empires of Asia, had masculine and feminine deities and forces reflected in the world around us. As far as I knew, the definition of "woman" was always presumed and didn't need much explaining. My first task was to figure out where the confusion entered in. When did people born men start thinking they were women, and people born women start thinking they were men?

IT'S IN THE BIBLE?

Of course, I had my theories that like Covid or Lyme's disease, the mass spread of transgender ideology began in a lab somewhere, probably in some weird university department devoted to a newly invented pseudoscientific "theory." But I wanted to keep an open mind. Even so, I was surprised when my first clue was dropped not from a historian, a priestess of an ancient religion, or a keeper of sacred texts, but instead from a gender surgeon who specializes in so-called "bottom surgery" (as I found out later, no, that does not mean surgery on the buttocks).

Dr. Marci Bowers was her name. Well, his name. The awkwardness of this whole transgender issue popped up rather quickly. Dr. Bowers was born as a male but surgically attempted to change his sex and now presents himself as a female. His long hair and feminized voice couldn't change his large facial features—or XY chromosomes for that matter. I wasn't there to discuss his particular gender journey, so when talking with him or about him with other people I interviewed, I just stuck with "you" or "Dr. Bowers." I will refer to everyone in this book by their biologically correct pronouns because it is more important to be grammatically correct than politically correct, especially when you're writing a book.

As a student of one of America's first gender reassignment surgeons, Dr. Stanley Biber, Dr. Bowers seemed like the perfect candidate to provide a firsthand look into the science behind gender reassignment and the developments in gender surgery—a subject I certainly planned to dive into with some detail. But when we started to talk, I soon realized that I had gotten more than I bargained for.

"It's part of human biology and human nature. Trans—it's probably been here since the beginning of time," Dr. Bowers told me as I sat in his medical clinic in Burlingame, California. Burlingame sits just south of San Francisco and north of Palo Alto at the heart of the California tech community. It's the intellectual center of the artificial intelligence-inspired philosophy of transhumanism—the idea that people, aided by technology, can transcend the constraints of the human species in its current form. It's a fitting place for a gender reassignment surgery clinic.

"Even in biblical times there are references to individuals who are probably trans," Dr. Bowers added.

I quickly interjected, "What, in the Bible?"

"Even in the Bible," he said.

I was anxious to learn more. I had read the Bible many times, and it always seemed pretty clear-cut on the gender issue. "Male and female He created them," as the author of Genesis wrote. It doesn't get simpler or more direct than that.[1] Not to mention modern day Christians tend to be the most thorough opponents of the transgender ideology. I listened further to Dr. Bowers, curious to hear his proof that gender theory can be traced back to the sacred texts of Jews and Christians. Perhaps he'd found a verse promoting transgenderism tucked away in some chapter of Scripture that no one else had ever thought to read.

"Even in the Bible there are passages, and there are clues that probably [transgenderism] was something that was happening at the time. There were things called eunuchs. There're 58 references to eunuchs, which are castrated males, which acts to feminize a person, just in the Bible alone. And the passages in the Bible like Matthew 19 say that adultery is expressly forbidden unless your husband is a eunuch."

It was news to me that the Bible allows adultery. The Sixth Commandment seemed to be rather straightforward on this point: "Thou shalt not commit adultery." Not to mention, I thought a eunuch was someone who had been castrated—not necessarily by their own will. Isn't that different than a transgender person? I had my doubts, but this was my first lead into where gender theory came from, so I had to see if the transgender doctor was right and everything that I had ever learned about the Bible was wrong.

I picked up my Bible and found out pretty quickly that the doctor was right about one thing. Matthew 19 did mention eunuchs. Unfortunately, he was wrong about everything else. The verses don't condone adultery, and they aren't even about transgenderism. The passage in question is somewhat mysterious. Jesus says, "For there are eunuchs who have been so from birth, and there are eunuchs who have been made eunuchs by men, and there are eunuchs who have made themselves eunuchs for the sake of the kingdom of heaven."[2]

If Dr. Bowers were right, that would seem to mean that men who castrate themselves are serving the kingdom of heaven. But that doesn't match with anything else in the Bible. I dove deeper in to see what some of the earliest Christian thinkers and scholars might have to say on the subject. Maybe they knew about transgenderism, and we somehow lost the knowledge and remained ignorant for centuries until Dr. Bowers came along to rediscover the long-hidden truth.

That's when I found some revealing passages from Origen of Alexandria, one of the most famous Christian theologians in history. He was alive in the 100s and 200s AD, so I figured if transgenderism really was a part of the early Church, he would know better than anyone.

I was lucky, because Origen wrote on that exact verse that Dr. Bowers mentioned. He wrote that God was talking about celibacy and "means not the excision of the members" in this verse. "Since the man who has mutilated himself, in fact, is subject even to a curse." It only got more extreme from there. "For to cut off our members has been from the beginning a work of demoniacal agency, and satanic device, that they may bring a bad report upon the work of God, that they may mar his living creature."[3] Looks like transgender doctors aren't exactly the best source on theology.

Perhaps Dr. Bowers misinterpreted that particular verse but was still right on the larger point. If Jesus Christ Himself didn't embrace gender theory, maybe it was present elsewhere in the Bible.

Deuteronomy 23:1 mentions eunuchs too, in a way. "He whose testicles are crushed or whose male member is cut off shall not enter the assembly of the Lord."[4] Alright, that must not be what Dr. Bowers was referring to either.

Later passages are a little kinder to eunuchs, but still don't dive into anything resembling gender theory. For example, the prophet Isaiah wrote, "For thus says the Lord: 'To the eunuchs who keep my sabbaths, who choose the things that please me and hold fast my covenant, I will give in my house and within my walls a monument and a name better than sons and daughters; I will give them an everlasting name which shall not be cut off.'"[5]

The Bible never actually mentions anything that we would recognize as the modern conception of transgenderism. It was a tenuous logic chain at best, and I soon realized that this entire argument has already been thoroughly disproven... in a pro-LGBT publication called *Grace and Lace*, nonetheless. The author in *Grace and Lace* minced no words: "The principal theme relating the transgender/transexual person to the 'eunuch' as described in the Bible is at best a questionable connection." The whole theory rests on conjecture that is "disputable and sometimes obviously incorrect."[6]

Eunuchs don't reference transgenders or transexuals or trans anything. In Biblical times, the term simply meant a castrated male. Some were castrated because they were prisoners or slaves, and they wanted to be deemed unthreatening to the women they served. Others chose to be castrated as part of masochistic pagan religious rites.[7] But there is no proof whatsoever that these men who were castrated or who castrated themselves believed they were women.

Another pro-trans author rejected the idea that transexuals would even consider themselves eunuchs, saying it is "Deeply offensive to reduce trans women to castrated men, which the term 'eunuchs' conventionally connotes."[8] Yet another author offered an incisive critique of the argument, calling the eunuch trope a method "employed by transgender Christians to alleviate [their] cognitive dissonance." Quite an indictment of Dr. Bowers.[9]

I was back to square one. On second thought, the Bible never did seem like the most auspicious place to locate the origins of gender theory. Attempting to do so was worse than trying to force a square peg into a round hole. It was attempting to make a square peg into a round hole by simply calling the square a circle. The assertions never matched the actual facts or text. Like all attempts to flip the Bible on its head, it appeared this was another way people were trying to coopt their opponents' arguments and use it against them. My search continued.

Dr. Michelle Forcier, an Associate Professor of Pediatrics and an Assistant Dean of Admission at the Brown University Alpert Medical School brought up my next lead. Generally, the longer someone's academic credentials, the less credible that person is. But Dr. Forcier does bill herself as a professional in sexual health issues for lesbian, gay, transgender, and queer youth. Maybe in her years of study she found the source of gender theory.

I asked her if she knew who first came up with the term "gender identity." "There's been two spirit genders in the American Indian culture," she said.

Interesting. I looked into the idea of "spirit genders" a bit more, and it won't surprise the reader to know that I found an entire article on the subject on the Biden Administration's Department of Health and Human Services (HHS) page. I was relieved to discover that the Administration is focusing on issues that really matter.

According to HHS, Native American "two-spirit people" combined the traits of both men and women but were considered neither male nor female. They also received their spirit (their gender?) from the gods in some sort of mystical way. Was this a common belief? Was it a big part of Native American culture? Does this have anything to do with a man becoming a woman or a woman becoming a man, or is it merely the spiritual idea that the gods can impart characteristics of both men and women?

Apparently, I wasn't the only one that thought rooting transgenderism in the idea of "two-spirit" was a bit thin. In none other than the Encyclopedia of Gender and Society, it says that the term "two-spirit" itself "emphasizes a Western idea that gender, sex, and sexuality are binaries. It implies that the individual is both male and female, and that these aspects are intertwined within them. The term moves away from traditional Native America/First Nations cultural identities and meanings of sexuality and gender variance." Furthermore, the encyclopedia noted that, "the idea of gender and sexuality variance being universally accepted among Native American/First Nations peoples has become romanticized."[10]

Apparently Dr. Forcier had decided that appropriating Native American history is acceptable when it serves transgender ideology. But it doesn't fit the bill for my purposes.

I turned back to Dr. Forcier and got at the question from a different angle. "What would you say to people who would argue that sex isn't so much just assigned by the doctors [at birth] as observed as a physical reality?" I asked.

"The reason goes back to, it was important to have gender or sex assigned on birth certificates because at one point in time, people who had a female gender assignment didn't have the same rights as men who had a male gender assignment. So that gender assignment on a birth certificate was important for who could own property and who was property." Dr. Forcier added, "[It's] an arbitrary social construction that continues with us today."

So doctors just arbitrarily "assigned" gender at birth because of property rights? In her reading, it seems that the main reason you say that someone is male or female is because you needed to decide who could own property, who couldn't, and who would be "owned" by others.

I guess the sex of the person didn't matter nearly as much to the doctors as the fact that, apparently, some arbitrarily chosen 50 percent of the population was supposed to own the land while another arbitrarily chosen 50 percent were supposed to be slaves. The fact that, in her reading of history, the 50 percent of the population that didn't have the same legal rights as others all had the exact same biological sex didn't seem to be a relevant factor. According to her, a midwife or a pediatrician would look at a child and say, "These [sex organ] parts are there, those parts are there, so now your child will be arbitrarily assigned this gender at this point in time or this sex at this point in time."

I decided to dive a little deeper. After all, if it's an "arbitrary social construction," then maybe transgenderism really has been with us since the dawn of time, and human beings have just been viciously repressing it for millennia.

"Those words are interchangeable, 'sex' and 'gender'?" I asked.

She parried the question. "I don't like to interchange them. I like to keep sex for the act of loving, the act of sexual behavior."

I kept pushing. "That's a sexual act. But I mean in terms of..."

She cut me off. "Those are parts that we can see, hormones that might make themselves evident, and internal organs that we can look for, chromosomes that we can assess and study... Then we know brain, of course, is all part of gender because their glands, as well as a cognitive and emotional function that has to do with gender that are all interrelated to all the pieces that go into like growing up and to male, female, non-binary, or another sort of identity of gender."

Wait, what? So sex is a biological fact? Or not? I was lost. Trying desperately to make sense of her seemingly incoherent ramblings, I tried to simplify as much as I could. "When the doctor sees the penis and says, 'This is a male,' as in 'sex of male'; that's an arbitrary distinction?"

"Calling that person 'male' is how we assign sex in the early years," she said. "Telling that family based on that little penis, that your child is absolutely one hundred percent male-identified no matter what else occurs in their life—that's not correct."

Yet again she started talking about identity and not if the word "male"—or the word "female" for that matter—reflected any actual biological reality. How can I understand where the theory of gender identity came from if the experts in gender identity can't distinguish gender identity from biological sex?

I attempted one last Hail Mary hoping for some clarity. "I'm also confused by the language, because I can't quite understand where you fall on the question of male and female and if that's a biological reality." I continued, "If I see a chicken laying eggs and I say, 'That's a female chicken laying eggs,' did I assign it female or am I just observing a physical reality?"

"Does a chicken have gender identity?" she retorted. "Does a chicken cry? Does a chicken commit suicide?"

Chickens don't, but the idea was starting to become more appealing to me the longer this conversation went on.

THE REAL ROOTS OF GENDER THEORY

It was beginning to dawn on me that the most prominent experts in the field of gender identity seemed to have no idea where the radical idea of gender identity came from. They can't even distinguish between "sex" and "gender," which presumably is the very root of their field. Like most revolutionaries, it didn't really matter to the likes of Dr. Bowers and Dr. Forcier where the idea of gender identity started. The past has no real bearing on the present, and what matters is that now we think that men and women can decide whatever they want about themselves.

But that wasn't satisfactory for me. At some point in history, there was a radical transformation in how people understood sexuality. At some point, the very idea of "gender" had to have been formed. It had to have started somewhere. So, I set out to find the answer myself. I soon came to realize that gender theory didn't begin in the Bible. It didn't start with the practices of native peoples or because of the legal delineations of property rights. Gender theory hasn't been with us at all times while being masked by "arbitrary social construction."

It was a much more recent invention than that, and its seeds were actually planted by a particular person, a German physician by the name of Magnus Hirschfeld. Few people in America have heard of him, but in the 1930s he was dubbed "The Einstein of Sex," and he's considered the primogenitor of the gay rights movement.[11]

Born in 1868 in what was then the Prussian Empire (now Poland), Magnus Hirschfeld spent most of his life in Germany where he became the world's most prominent so-called sexologist in the early years of the 20th century.[12] A physician by trade and a homosexual by persuasion,[13] Hirschfeld travelled to Chicago shortly after medical school where he immediately began exploring gay subcultures.[14]

Upon his return to Germany, he founded what is considered the first ever gay and trans rights organizations called the Scientific-Humanitarian Committee.[15] Long before the ideas of the LGBT movement were popularized—or even tolerable to the larger public—he proposed that same-sex attracted people were not only born that way, but also meant to be that way. "Homosexuality was part of the plan of nature and creation just like normal love," Hirschfeld said in a 1907 testimony defending an army officer accused of having gay sex.[16] Suffice it to say, Hirschfeld was very, very ahead of his time.

By 1919, Hirschfeld had founded his Institute for Sexual Science in Berlin, a one-stop shop for counseling, political advocacy, public education, and research on gay issues.[17] In a way, it was the Human Rights Council of that era. Nine years later, he initiated the World League for Sexual Reform,[18] which hosted international conferences. Among the league's aims were "liberation of the marital relationship from Church domination," "application of the knowledge of Eugenics towards improvement of the race through Birth Selection," "proper, scientific understanding of variations in sexual constitutions (intersexuality)," and to free sex from being "complicated by any sense of guilt."[19] To put it more succinctly, the League wanted to overturn traditional ideas of sexual morality, marriage, and the relationship between men and women.

Back at that time, people didn't have the infinite variety of gender categories we do now. People didn't even really have an idea of what "gender" was. Hirschfeld broke ground by coining new terms like "transvestitism,"[20] which he distinguished from homosexuality,[21] as people who proactively hoped to change their sex.[22] In fact, Hirschfield described a multiplicity of what he called sexual "intermediaries" like hermaphrodites, androgynes, homosexual, and transvestites—all of whom were deemed a "third sex" who deviated from the male and female norm.[23]

At the same time, some of his comments hinted at the idea of gender fluidity that would be explored in much greater detail decades later. When speaking about women's rights, Hirschfeld is reported to have said that "the woman who needs to be liberated most is the woman in every man, and the man who needs to be liberated most is the man in every woman."[24] Above all others, Hirschfeld was taking the first, tepid steps of forming the idea of gender as distinct from sex.[25]

Doctors affiliated with his Institute for Sexual Science performed some of the earliest known sex change surgeries on these so-called transvestites.[26] These developments found some popular support in interwar Weimar Germany, where Hirschfeld "considered [it] his greatest triumph, for... the German government fully accepted and supported his theories."[27]

While news reports of these harrowing surgeries did make it into the American press in the 1930s, the popularization of gender transitioning took decades to actually take hold in the American consciousness.[28] That wasn't for lack of trying. Hirschfeld travelled across the world, including to Japan, China, India, Egypt, and the United States, among other countries.[29]

Hirschfeld's research and surgical experiments were quickly ended, however, when Hitler rose to power in 1933. The Nazis soon destroyed his Institute and burned his files, and Hirschfeld died in 1935, exiled in France.[30]

Hirschfeld is the grandfather of the modern LGBT movement. Professor Dagmar Herzog of the City University of New York

Graduate Center underscored Hirschfeld's influence: "[He] was 'the first' in so many ways: the founder of the first gay and lesbian rights movement, launcher of the first campaign to decriminalize homosexuality, first vocal and empathetic defender of transgender rights (including facilitating early gender confirmation surgeries), first to open an institute of sexual science, first to start a medical journal dedicated to sexual minorities, first to use film and pamphlet literature and public talks to combat popular anti-homosexual prejudice, first to develop support groups for same-sex-desiring individuals in order to facilitate self-acceptance."[31]

The question remained how these novel and subversive theories crossed the Atlantic and made it into America. That was facilitated by another early sexologist named Harry Benjamin. Benjamin was born in Berlin in 1895, twenty-seven years after Hirschfeld, and accompanied Hirschfeld in his exploration of gay subcultures in the German capital city.[32] By 1913, Benjamin was an expert in endocrinology—or the human hormone system—and had relocated to New York, though he visited Hirschfeld and his Institute for Sexual Sciences regularly in the 1920s and 30s.[33] He even managed one of Hirschfeld's American trips where the "Einstein of Sex" continued to research his theories stateside.

Benjamin began offering hormone therapies to cross-dressers and others who desired to live differently from the sex they were born, but at that time sex change therapy was far from common.[34] While Benjamin did act upon these nascent gender theories, he did not popularize them in America. In fact, he never even wrote on the subject until 1953 when he stated, "sex is never one hundred percent 'male' or 'female',", and attributed the idea of "intersexes" to both psychology and more natural causes.[35] Truth be told, in the 1930s, 40s, and 50s, America was hardly a fertile place for gender theory—certainly not as open to the ideas as Weimar Germany.

For these ideas to take root, the American idea of sexuality had to be fundamentally overturned. Hirschfeld and Benjamin may have been developing the doctrine and honing the message, but they needed a voice crying out in the wilderness to prepare the way of transgenderism, to make queer the straight path, and to condition people for this new "truth." That man was Alfred Kinsey—a personal colleague of Harry Benjamin who received materials from Benjamin for his research and who, in return, referred patients to Benjamin for therapy.[36]

THE SEXUALLY CURIOUS CASE OF ALFRED KINSEY

Lucky for me, one of the experts on transgenderism and gender theory that I had reached out to seemed to know quite a lot about Alfred Kinsey. Dr. Miriam Grossman, a certified child, adolescent, and adult psychiatrist, has researched the history of transgenderism and sex education, and now she has a mission to defend people—particularly children—from the dangers of this movement.

I met Dr. Grossman at her home office in New York, and we sat down for an extremely wide-ranging and fascinating discussion. According to Dr. Grossman, the dam holding back gender theory began to break in the middle of the 20th century and, like so much social degradation, it started by targeting children. Before children were ever injected with opposite sex hormones or told that they could be a man born in a woman's body or a woman born in a man's body, there first had to be a disruption in the understanding of childhood and development and of sex itself. Kinsey was just the man for the job.

"Kinsey was a social reformer; he wanted more than anything to change society," Dr. Grossman told me. "He wanted to rid society of Judeo-Christian values when it came to sexuality, and he worked very hard to do that. And I would say he succeeded."

Born in 1894 in Hoboken, New Jersey,[37] Kinsey was the shy son of a domineering father who went on to study biology at Bowdoin College in 1916. According to his biographers, his younger years were marked by sexual frustration. Like Hirschfeld, Kinsey was a homosexual, and he resented his Methodist upbringing that "repressed" his desires.[38] He soon was granted a professorship in zoology at Indiana University in Bloomington, where he later began his study of sex.

Previously when people studied the physical aspects of sex (if they studied sex at all), they researched the transmission of venereal diseases. No one before Kinsey had attempted to research and describe actual sexual practices among Americans.[39] In this approach—an approach pioneered by the likes of Hirschfeld and Benjamin—sex was no longer governed by the dictates of morals or truth, but by the concepts of health and freedom. It was a seismic intellectual shift, and it had profound consequences not only on America, but the world.

Kinsey's project was rooted in the assertion that people are sexual beings "from cradle to grave" as Dr. Grossman put it. This includes children, of course. As such, the sexual nature of children, in Kinsey's view, needs to be affirmed and cultivated. "[According to Kinsey], when children and teenagers are repressed by [Judeo-Christian] values, that is when people begin to suffer terribly," Grossman said.

Kinsey didn't exactly have the academic background to make such an assertion. He wasn't a physician or a psychologist. In fact, he was a zoologist whose expertise was in wasps—and he studied wasps intensively for roughly eighteen years[40] before shifting his focus to sex in 1938. It wouldn't be the last time that an ideologue would use pseudoscience to try and undermine the sexual mores of American culture. Kinsey's success inspired many imitators.

It all began rather serendipitously when Kinsey was asked by his university to chair a faculty committee that would design a course on sexuality and marriage. At that time, he was seen as a respectable and scientifically rigorous man, and he took his approach to the meticulous collecting and cataloguing of wasps into the field of sex.

Yet from the outset, there were hints that Kinsey wasn't merely going to "follow the science." He had an agenda. According to author T.C. Boyle, "Kinsey electrified the assembled students by announcing at the outset that there were only three types of sexual abnormality—abstinence, celibacy, and delayed marriage—and he absolutely stunned them by showing slides of sexual intercourse."[41] Even before his course on marriage started, and he was still just a kooky wasp collector, Kinsey was asking his students about their sex lives.[42] The questions he posed to those under his tutelage were increasingly specific, such as when they first had premarital sex, how often they had sex, and the numbers of partners they had. People may talk about those subjects freely and proudly these days, but in the 1930s it was unheard of. Where supporters of Kinsey saw academic openness, opponents saw voyeurism.

As the years went on, Kinsey began collecting more and more data in order to generate statistically significant findings, allegedly going through great lengths to ensure that his interviews produced the most accurate results possible.[43] In a way, Kinsey followed Hirschfeld's lead, using questionnaires and interviews to research sexual behavior in America the same as Hirschfeld had done in Germany.[44] Over the course of his life, Kinsey personally interviewed 8,000 people, and he and his team together had interviewed 18,000 people. As a testament to elite support for his nascent project even in the pre-Sexual Revolution times, Kinsey received funding from the quintessentially establishment Rockefeller Foundation to continue his work.[45]

Over the years, Kinsey would follow up with his interviewees, asking for them to send their friends or other people they knew to him to be interviewed. Coincidentally, this chain of interviews at one time led him to the very same Chicago gay subculture where Hirschfeld had started his sex research decades before.[46]

After years of work, Kinsey's first report entitled *Sexual Behavior in the Human Male* was published in 1948, followed by *Sexual Behavior in the Human Female* in 1953.

"His research was published to great fanfare," Dr. Grossman told me, "because in his research he claimed that... the way that we think of how people are living in terms of their sexuality is just a complete cover-up." Standing alongside researchers and professors in white coats and armed with diagrams and statistics, Kinsey allegedly revealed that everyone was living a double life. "Almost everyone in the country, male and female, is living a life of sexual experimentation, freedom, multiple partners, different sexes, different activities, children, babies... and he claimed that his research proved it," Dr. Grossman said. According to one biographer, his findings were "being discussed in homes, grocery stores, break rooms, and on the radio,"[47] and another said that his major accomplishment "was to challenge most of the assumptions about sexual activity in the United States."[48] The wall guarding how, when, and where society talked about sex was beginning to crumble.

Kinsey developed a scale—later dubbed the Kinsey Scale—that sought to measure a man's homosexual tendencies on a single continuum, asserting the idea that sexuality is something fluid and changeable over the course of one's life.[49] A zero on the scale meant that a man has no homosexual attractions whatsoever; a six on the scale meant he had no heterosexual attractions what-soever.[50] A three on the scale presumably meant the man was what we now call a bisexual, though that was a category Kinsey personally rejected.[51]

Kinsey proposed that his research denied basic assumptions about marriage. For example, he questioned whether adultery actually undermined marriage itself and argued that an appro-priate extramarital affair is one where neither party becomes emotionally involved.[52] The idea is not only laughable, but easily recognizable as dangerous to any human being who has ever had a successful marriage. But on the surface, it seemed like the "science" had spoken. As a further testament to Kinsey's divorce from reality, in all of his studies and interviews and questioning, he apparently showed almost no interest in the primary bio-logical end of sex: pregnancy.[53]

Central to Kinsey's impact was the idea that he had allegedly interviewed thousands of normal, everyday Americans and asked thorough questions about their sexual practices.[54] His bi-ographers argue that he started by interviewing college-educated, middle- and upper-middle class individuals, followed by lower class people.[55] Regardless, Kinsey implied that his research was focused on the most normal and respectable of people in the nation. The self-professed goal was to describe not what normal

Americans were supposed to be doing in the bedroom according to the old social norms, but what they were actually doing.

And what normal people were actually doing, according to Kinsey, was shocking. People started to wonder if dear old average Bob and Susan down the street were, statistically speaking, complete nymphomaniacs, and maybe strange sexual practices really weren't so strange after all. If as many people acted like this as Kinsey said and society was still up and running and, well, normal, then perhaps all the taboos and rules that appeared to govern sex really were just meaningless and hypocritical restrictions.

Except that it was all a lie. It took years, but a researcher named Judith Reisman uncovered the truth. "[Kinsey] was interviewing convicted sex offenders," Dr. Grossman told me. "He was going into jails and interviewing child molesters, people who committed sexual assault. He was interviewing prostitutes...terrible experiments [were] done on children under year one. They were basically being sexually assaulted." According to Dr. Grossman, Kinsey's entire methodology was bunk.

Additionally, Kinsey's focus on perverted sexual acts stemmed from the same force: Kinsey himself was the sexual deviant living the very life he said every normal American was living. This isn't conjecture. It was even reported in the *New York Times*: "Kinsey had had affairs with men, encouraged open marriages among his staff, stimulated himself with urethral insertion and ropes, and filmed sex in his attic."[56]

Dr. Grossman's theory was that Kinsey was projecting to rationalize his own perversions. "What he wanted to do is to be able to say, 'No, it's not like it's just me. Everyone's like this. Everyone's like this, not me.' I mean, I'm a psychiatrist, but it's pretty clear what his motivation was."

KINSEY'S CREEPY FOCUS ON KIDS

Most disturbing of all was Kinsey's research on child sexuality. According to Reisman and the Child Protection Institute, Kinsey's 1948 book on male sexuality included five tables of data on "multiple orgasms in pre-adolescent males." What qualifies as a "pre-adolescent male"? The table makes it clear by listing infants as young as two months up to boys the age of fourteen, noting how many "orgasms" these young boys were "observed" to have had.[57]

This data was taken from adult pedophiles who had "sexual contacts with younger boys" and who were able to "recognize and interpret the boy's experiences."[58] Reisman filled in the details, noting that child sex abusers "used stopwatches and took meticulous notes that were transmitted to Kinsey."[59]

However, it wasn't just the number of orgasms Kinsey noted. In cold and clinical language, he described what happened to some of these boys. "Extreme tension with violent convulsions: Often involving the sudden heaving and jerking of the whole body. Descriptions supplied by several subjects indicate that the legs often become rigid, with muscles knotted and toes pointed... eyes staring or tightly closed, hands grasping, mouth distorted... whole body or parts of it spasmodically twitching... Sobbing, or more violent cries, sometimes with an abundance of tears (especially among younger children)... in some individuals involving several minutes (in one case up to five minutes) of recurrent spasms."[60]

While Kinsey always proposed that he was merely describing reality, not passing judgment, and while theoretically he could have not requested this data and merely received it totally unsolicited (both big "ifs"), this research details nothing less than child rape. Whatever the case, Kinsey held that so-called inter-generational sex—a polite euphemism for child rape—poses no serious harm to children.[61]

Ultimately, his abuse of children wasn't a bug of Kinsey's research, but a feature. "This also fits with that agenda because if children are sexual and they enjoy sexual touch and sexual activities, then, hey, what's wrong with an adult engaging in those activities?" Dr. Grossman said. As Kinsey biographer Jonathan Gathorne-Hardy put it, "Theoretically, therefore, as far as Kinsey was concerned, there was nothing automatically wrong with child-adult sex."[62] Like the most extreme of contemporary sexual ideologues, the only guard against pedophilia was the idea of "consent," though that begs the question of how a child could ever actually consent to an adult's sexual advances.[63]

I asked Dr. Grossman if there was anything at all scientifically valid about Kinsey's research. "Maybe from his research about wasps," she retorted. "I think in this area [of sexuality] he was a fraud."

Unfortunately, these lies were only revealed much later. In the meantime, Kinsey was dubbed "The Father of the Sexual Revolution," and elite consensus had adopted Kinsey's findings as fact and integrated them as a core component of their sexual ideology. His theories about child sexuality began to completely unmoor traditional ideas about development, gender roles, and human attraction. The abusive and perverted actions of a sick pedophile created the aura that absolutely anything goes when it comes to sex.[64]

And if children were sexual beings, and, according to Kinsey's research, everyone engages in all types of diverse sexual practices, who's to say that children only desire or engage in heterosexual practices?

Adults with their own peculiar sexual fantasies and tastes began to project their desires and confusions onto children. It was only a small jump from that to the supposition that children are born with a free-floating gender too. Every idea our culture had about sex was being questioned and overturned. The old rules didn't apply anymore. It was a brave new world, and ideas and theories that would have been rejected outright before suddenly became possible.

While the door was beginning to open to the idea of transgenderism, Kinsey himself didn't approve of genital surgery in an attempt to change one's sex. He wrote that "a male cannot be transformed into a female through any known surgical means. In other words, it would be very hopeless to attempt to amputate your male organs and implant a vagina."[65] Of course, such a statement would get Kinsey immediately cancelled today, but back then the exploration of so-called transvestitism was so new that he has since been given a pass.

Despite his rejection of gender change surgery, he did take a keen interest in what would become "transgenderism." As he was publishing his two works on the sexual behavior of males and females, he also began moving beyond the study of heterosexual and homosexuals and looked into those who practiced cross-dressing and identified as cross-gendered.[66] His research on the cross-gendered didn't appear on those two seminal works, so its popular influence was not immediately felt.[67] Nonetheless, Kinsey's research seemed to be following a natural development.

By all accounts, even Kinsey, transgressive as he was, didn't consider so-called cross-gender people to be in their own category and certainly didn't think it was an exceedingly common phenomenon.[68] By the early 1950s, Kinsey turned much more attention to the subject of transvestites, and he worked with a transvestite named Louise Lawrence who connected him with a small network of cross dressers across the country whom he could interview.[69]

Though he eventually interviewed one hundred men who considered themselves women, another eleven women who considered themselves men, and another ten men who had actually undergone sex change operations in an attempt to become women, he had only enough material by 1953 to publish a few paragraphs on the subject of transvestites in his second work, *Sexual Behavior of the Human Female*. One detail he did note was that, at that time, "an exceedingly large proportion of transvestites are anatomically males who wish to assume the role of females." He estimated that only two to six percent of transvestites were anatomical females.[70] This is directly contrary to current trends where young women are embracing transgender identity in droves, but we'll get into that later.

Kinsey died in 1956 before he could expand his research on transvestites to the same degree that he had studied male and female sexual behavior—so the baton he ran with had to be taken up by someone else.

As historian and sexologist Vern Bullough wrote in the *Journal of Sex Research*, Kinsey is in "the pantheon of pioneer researchers" of sexology in the 20th century. Where Kinsey "described the varieties of sexual behavior of Americans," another man "took the next step and constructed a theory of sexual development, emphasizing the interaction and interdependence of social, psychological, and biological factors."[71] That man wove together the disparate threads and earliest hypothesis of gender theory into a more cohesive whole. Effectively, he was the father of what we now recognize as gender theory, and his name was John Money.

JOHN MONEY

BEFORE JOHN MONEY, there was a small group of people slouching their way towards gender theory and transgenderism—people like Magnus Hirschfeld and Alfred Kinsey—but it hadn't taken the shape of gender theory as we recognize it today. John Money was the one who first put the pieces together. In fact, he was the first person to use "gender" as a term distinguishable from sex.

But as I was conducting my interviews, few people brought him up. Did they not want to talk about him? Did they not know about him? I was desperate for more answers, so I dug deeper.

I decided to start where I had left off.

FOLLOWING MONEY

It seems John Money had quite a few links to Alfred Kinsey. Money was considered one of the core researchers at Kinsey's "Institute for Sex Research" (renamed the Kinsey Institute for Sex Research after Kinsey died).[1] Money was actually a mentor to June Reinish,[2] the second director of the Kinsey Institute after Kinsey's death.[3] Money's archives—including his manuscripts, articles, media interviews, lectures, correspondence, and more—are all housed at the Kinsey Institute Library.[4]

John Money also had connections to Harry Benjamin—one of the people who served as a link between Hirschfeld and Kinsey.

Money was part of Harry Benjamin's research team on transsexualism from 1964 to 1967,[5] and he worked closely with the Harry Benjamin Foundation to receive patient referrals.[6]

That being said, Money may not have been the biggest personal fan of Kinsey. At one time, Money said, "The cumulative effect of Kinsey's way of communicating with people could not have been better calculated to antagonize."[7] At the very least, Money and Kinsey had professional ties and ran in the same circles, which should be no surprise considering how small the circle of mid-20th century libertine sexologists and wasp collectors must have been.

When I brought up the subject of John Money to Dr. Forcier, she took a defensive and somewhat dismissive posture. "[John Money] is someone who did early gender work, along with what we've learned about gender from the intersex community," Dr. Forcier told me, "which has been hugely powerful." Others described him as a "truly original thinker," a "pioneer in the truest sense," and one who had a "passionate commitment to the rights of the individual." Does this mean that he wasn't a perverted psychopath like Kinsey? Let's follow Money's story and find out what he did to become the "Father of Gender Theory."[8]

Born in New Zealand on July 8, 1921, Money was another sexologist allegedly marred by a troubled childhood.[9] "He was raised by a violent, alcoholic father who would regularly beat [Money] and his mother," Dr. Grossman told me. "So he witnessed domestic violence to a severe, severe degree." Dr. Grossman speculated that as a result, Money developed what would now be called "gender dysphoria"—a severe uncomfortableness with one's sex. "His primary masculine role model was a monster. He would hang out much more with the females in his family and his relatives, because this man terrorized everybody," Dr. Grossman explained. "Now, Money wrote that he suffered—he bore the vile mark of being male. It's not a direct quote." I looked up what Money had said: "I suffered the guilt of being male. I wore the mark of man's vile sexuality."[10]

It's not exactly a surprise, then, that Money was only married once in the 1950s, and it soon ended in divorce.[11]

As a psychiatrist, Dr. Grossman saw a common thread: "You can see how... if you look at the psychology of both John Money and Alfred Kinsey, you see the suffering that they went through and how they came up with theories that would give them some relief from that suffering."

After graduating from Victoria University, Wellington, with a teacher's certification and two master's degrees—one in education and another in philosophy and psychology—Money worked a few years in the psychology department of the University of Otago in Dunedin.[12] It was only a brief stint, however, because by 1947 Money had immigrated to the United States, eventually entering a PhD program at Harvard University.[13] The subject of his dissertation was the core interest of his life's work. It was entitled "Hermaphroditism: An Inquiry into the Nature of a Human Paradox."[14]

The goal was to study something called intersex, a condition, as Dr. Grossman explained, "which means both male and female sex organs or parts of them exist within one person. It used to be called hermaphroditism; it's now called intersex." This was a legitimate course of study, as some babies are actually born with this condition. Yet according to Dr. Grossman, it's an extremely small number. "Intersex babies are born in about one in ten thousand births," she told me. From the existence of that small population, Money established an extremely bold hypothesis.

"John Money came up with a theory that each of us is born a hermaphrodite, at least psychologically," Dr. Grossman told me. Effectively, nature means nothing, and nurture is the only aspect that influences whether someone grows up to think of him or herself as a male or female. This fits in very nicely with modern theories about gender roles—not to mention the very ideas of masculinity and femininity—as all being "social constructs." If nurture is everything, then there's no real, biological reason whatsoever that a girl would prefer to play with dolls and a boy would want to play with trucks.

Dr. Grossman conjectured on why Money would pursue this theory so doggedly. "When he came up with his theory, when John Money came up with a theory that each of us is born a hermaphrodite; we, at least psychologically, could be either. To him, that would have been an answer to his distress and his pain, not wanting to identify as a male, as a father, as his father."

Though Dr. Grossman obviously disagrees with Money, she said that Money's theory was slightly more credible when it was first proposed. "At the time when John Money came up with that [his theory that only nurture counts and not nature] you could almost believe it," she said, "because we couldn't look at the X and Y chromosomes the way that we can now."

It's important to note that while Money's theories have obvious implications for science, medicine, and society, Money himself was a psychologist—he wasn't a medical doctor or a surgeon or even a biologist.[15] In a way, his theory was divorced from biology because, by training, he focused on behavioral aspects alone.

"In intersexed individuals," wrote Joanne Meyerowitz in her history of transsexuality in America, *How Sex Changed*, "the sense of being a man or a woman resulted not from hormones, gonads, chromosomes, or other physical variables, they [Money and his fellow researchers, Joan and John Hampson] argued, but from the sex to which the infant was assigned and in which the child was subsequently reared." Things like sex hormones, wrote Money, "have no direct effect on the direction or content of inclination. They are assumed to be experientially determined."[16]

As the *New York Times* wrote in 1973 while discussing Money's "findings." "If you tell a boy he is a girl, and raise him as woman, he will want to do feminine things."[17]

As a result, Money and his team chose the word "gender" as something psychologically distinguishable from sex. Money defined sex by six different factors: assigned sex, external genitalia, internal reproductive structure, hormonal and secondary sex characteristics, gonadal sex, and chromosomal sex.[18] His distinctions are still in use today and have been expanded upon. Dr. Marci Bowers, the sex change surgeon in California I spoke with, used Money's language almost word for word, telling me that, "there are at least a dozen different biological measures of sex. There is chromosomal sex, there's hormonal sex, there's anatomical sex."

In Money and Bowers's system, all these categories of biological sex are utterly distinct from gender. Meyerowitz wrote that "in 1955 Money used the term 'gender role' to refer to 'all those things that a person says or does to disclose himself or herself as having the status of boy or man, girl or woman' and 'gender' to refer to 'outlook, demeanor, and orientation.'"[19] Ultimately, the sex was the biology, and the gender was everything else—and the two didn't need to match.

It's not that "gender" is a made-up word. It was used before, predominantly as a way to describe whether words in other languages were masculine or feminine. Ultimately "gender" has its roots in the words (ironically) "gene," "genus," and "genre"—all ways of classifying and ordering or grouping people or animals or objects together.[20] So how did this linguistic and organizational term come to take on a nearly completely different meaning—to refer to something that is being used, at least nowadays, as a tool to destroy science-based distinctions and along with it our ability to order or understand anything—to include the definition of the word "woman"?

I'll let John Money speak for himself in his admittedly, less-than-clear style. "Because sex differences are not only genitally sexual, although they may be secondarily derived from the procreative organs, I found some need thirty years ago for a word to classify them. That word, which has now become accepted into the language, is gender."[21] Ironically, Money, by an apparent act of will, chose a word meant to order and changed it to create confusion and disorder. It wouldn't be the first or the last time in my journey that I discovered proponents of gender theory abusing and misappropriating language.

Dr. Grossman dove into Money's theory further, saying that "the idea that [Money] introduced to the world and that he worked his entire life to promote and to prove was that when a baby is born, it's gender neutral. When a baby is born, it has no identity as male or female. That develops in the first years of life, depending on what the message is that the child gets from the parents, from school, from society." According to this hypothesis, whether one acts as a man or as a woman (whatever those terms mean) was completely determined not by biology and not even by psychology, but by rearing alone.

If that's the case, then what does it mean when someone acts masculine or feminine? Just as with gender, Money unmoored gender roles from any attachment with physical reality or biology. These roles were all, to use Dr. Forcier's words, "arbitrary social construction." She wasn't the only one to speak that way. Dr. Bowers also told me explicitly that "gender and gender identity is really a social construct. It's the cues that you give." Dr. Bowers went further, saying that sex change surgery "never does make a person male or female" not because changing sex is impossible, but rather because sex is "more of a societal construct." The proponents of gender theory were all reading from the same script, and that script was written by John Money.

Meyerowitz drew a direct connection between this theory and Kinsey: "Alfred Kinsey and his colleagues had rejected both the theory of human bisexuality and psychoanalytic theories of early personality development. They argued that much of human sexual behavior, including homosexuality and transvestism, resulted from 'learning and conditioning.' Similarly, in their studies of intersexed conditions, Money, Hampson, and Hampson pointed to forms of social learning as the source of gender."[22] One man was building upon the work of another.

In Money's early works, he didn't appear to sever the idea of gender from sex in an attempt to subvert traditional masculine and feminine roles in society, or as an excuse to undermine people's understanding of their own sex. In fact, John Money wrote in one of his early studies entitled "Incongruous Gender Roles" that to successfully rear a child, you should be "orienting him, from birth, to his biologically and culturally acceptable gender role" and further, that his parents should "exemplify these respective roles" as well.[23]

But if these roles were truly arbitrary, it's hardly a shock that what counts as a "biologically and culturally acceptable gender role" would soon be subverted. Unsurprisingly, there is a popular idea today that children are not only blank slates when born, at least in regards to gender, but should also be raised in an androgynous environment and urged by their parents to choose their own gender—and continue to cycle through an infinite array of gender options as they develop. That idea can be clearly traced back to John Money.

"Gender... is a perception. It's a feeling," Dr. Grossman told me, describing how Money's theory played out. "It's a way of identifying. It's an experience. Ok, that's subjective. That's up to the person. That can change from day to day."

Money's supposed finding had such a massive impact on society that journalist and author James Lincoln Collier wrote in the *New York Times* that John Money's research "is the most important volume in the social sciences to appear since the Kinsey reports."[24]

Money's theories gradually evolved. For example, while at first claiming that nurture alone was the defining factor of whether one acted masculine or feminine, he later developed the idea that prenatal hormones can influence behavior as well,[25] opening the door to modern experiments with hormone therapy. However, his baseline assertions remained that gender is different from sex, and that gender is much more pliable than people had ever dreamed before.

MONEY'S POLITICAL (AND SEXUAL) AGENDA

Typical of so many popular scientists, Money was hardly a dispassionate observer of research, data, and the facts. Like Hirshfeld and Kinsey, he had a political agenda. Meyerowitz wrote that "John Money saw himself as a pioneer who rejected the vestiges of Victorian prudishness and championed sexual liberty. Some people saw him as brilliant and charismatic, and others disliked him with a certain intensity. He was confident to the point of arrogance."[26]

That's hardly the disposition of a scientist. Of course, how one is raised and what gender roles exist help determine the ways in which men and women act within a society. But could any reasonable person really claim that there's nothing innate at all about being a woman? Are human beings just an arbitrary and bizarre combination of learning and conditioning, with some hormones thrown in?

The evidence began piling up that Money was really driven by a sort of Promethean hubris. At first, I noticed it in how he treated language. Though Money claimed to be inspired by Ernest Hemingway, praising his "economy of words and uncluttered style,"[27] he not only coined terms when he thought it was necessary, but changed formerly clarifying terms in ways that would fit his narrative. This went beyond creating the term "gender."

He altered the phrase "sexual preference," which was used more commonly in the past to refer to whom one prefers to have sex with, and changed it to "sexual orientation," implying that one really has no choice over the matter, one is simply directed a certain way from birth.

Several of Money's other terms didn't make it into the popular dictionary, but nonetheless reveal the way he thought. For example, he attempted to change the word "perversions" to "paraphilia"[28]—a Greek combination word that combines "love" (philia) with "beyond" or "out" (para). The idea was that sexual perversions—the stuff that most people find degrading or offensive, like sadomasochism—really aren't gross or even abnormal.

They are just expressions of "love" that operate "beyond" what polite society considers respectable. While Money's term never did catch on, the idea behind it obviously has, as we shall see.

Soon I found Money's most disturbing alteration of the dictionary. In his book *The Adam Principle*, Money distinguished between "pedophilic sadism"[29] and "affectional pedophilia,"[30] which presumably means when the younger partner to the sexual interaction has affection for the older partner—as if somehow that makes an older person preying on a younger person acceptable.

Money discussed pedophilia further in his book without condemning it—as long as it didn't descend into violence. In fact, he wrote positively that pedophiles and the children they abused can even have a sort of happily-ever-after experience. "For both partners in a pedophilic relationship," Money wrote, "the break in their erstwhile erotosexual interaction [which happens when the young person is too old to arouse the older pedophile] does not dictate a complete estrangement, but allows the continuance of nonerotosexual friendship."[31]

Considering his seemingly pro-pedophilia comments, it's no wonder that Money decried how "the legal age of childhood was raised by act of Congress (Public Law 98-292, the Child Protection Act of 1984) so as to facilitate and expand prosecution in cases of pedophilia."[32]

Got that? Money thought Congress raised the age of what we considered children just so we could have more prosecutions—not to protect young people from the predatory attacks of adults.

I thought it couldn't get worse, but I soon found that Money wasn't merely trying to distinguish different *types* of pedophilia, ostensibly to make some forms seem more acceptable than others.

Nor was he hiding his support for the idea that young girls and boys can have sexual relationships with older men and women without any negative impacts. At one point, he made it explicitly clear that he supported the sexual abuse of minors. "If I were to see the case of *a boy aged ten or eleven*," Money said (emphasis added), "who's intensely erotically attracted toward a man in his twenties or thirties, if the relationship is totally mutual, and the bonding is genuinely totally mutual, then I would not call it pathological in any way."[33] Of course, it's impossible for a boy of ten or eleven to have a mutual bond with a 35-year-old male. Boys that age are incapable of it physically and emotionally, and they shouldn't be subject to the lust of men three times their age.

Money's support for pedophilia goes far beyond an exercise in dismantling Victorian prudishness. This is the full-scale destruction of any moral guardrails whatsoever, and the end result is nothing less than throwing the most vulnerable and defenseless people in our society—our children—to the wolves, subject to the sick and twisted desires of those around them. Oh, and it all comes with the approving stamp of "science" from one of the world's foremost sexologists, after all.

It was becoming clear that full and complete sexual anarchy isn't a corruption of the project of the transgender movement or of the sexual revolution, for that matter. In fact, the exercise of unrestricted desire is at the core of the project of sexual liberation—and transgenderism is just one part of this movement.

All of a sudden, my mind jumped back to Alfred Kinsey. Who was it that exposed Kinsey again? Yes, it was Judith Riesman. I pulled out her research and found something I had missed in the footnotes before. It was a story about how Reisman introduced her findings to the Fifth World Congress of Sexology in Jerusalem in 1981.[34] She wrote that at that conference, she presented a paper entitled "The Scientist as Contributing Agent to Child Sexual Abuse: A Preliminary Consideration of Possible Ethics Violation."

There, Reisman presented on Kinsey's child sexual experiments—including the horrifying details of so-called "pre-adolescent orgasms." She showed slides of the tables from Kinsey's book to a standing-room-only audience, and at the end of her presentation, everyone stood in stunned silence.

That's when a Swedish reporter jumped in. According to Reisman, he "cut into the anxious room. He declared to those present, all leaders of the human sexuality 'field' from many countries including England, Norway, Sweden, Denmark, Ireland, France, Canada, Germany, and the United States, that this revelation on Kinsey's involvement with children is an 'atomic bomb,' and he demanded to know how they could just sit there."

John Money gave the keynote address at the very same conference, and he heard what was happening. This is when he could have spoken clearly in defense of children. He might have decried any instances of sex abuse of minors, knowing that no worthy moral system could ever condone abuse for the sake of alleged scientific gain. He could have done that or any number of things. Instead, Money pushed open the doors to the main conference room, walked onto the podium, took a microphone from the moderator, and spoke to the anxious crowd.

"[If] this woman... is allowed to continue, sexology and sex education will be set back 200 years," John Money said, according to Riesman. He said that their livelihoods as so-called "students of sex" depended on not letting the name of Kinsey—the "Father of the Sexual Revolution"—be tarnished. John Money defended pedophilia. He tried to erase the idea that there could be such a thing as a sexual perversion. And right there in Jerusalem in 1981, in front of a packed room, he openly defended child sexual abuse for the sake of protecting sex research.

THE RISE OF SEX CHANGE OPERATIONS

Money clearly was determined to normalize sexual conduct that had once been beyond the pale. His most notable success in this arena was to take something that would have previously been considered an unthinkable Frankenstein project and move it into the mainstream: sex change operations.

John Money wasn't the first to conduct a sex change operation, but he may have done the most to make the idea palatable to polite society. It started with his association with the prestigious Johns Hopkins Hospital. Money started at Johns Hopkins Hospital and School of Medicine in 1951, and he remained a Professor Emeritus of Medical Psychology and a Professor Emeritus of Pediatrics there for over a half century until his death in 2006.

By the mid 1960s, Johns Hopkins had hosted John Money and his fellow researchers looking into the subject of transsexualism for years, and the school and hospital had been considered a center for the study and treatment of intersex people for decades.[35] That and the fact that the sexual revolution was well under way apparently made the idea of sex change surgeries seem more acceptable to the hospital.[36]

John Money—who, as a reminder, was not a doctor—and his team of doctors saw their first surgical patient in 1965. His name was Avon Wilson, an African-American man who thought himself a woman. Harry Benjamin helped make the referrals, and he said that Mr. Wilson's genital surgery would "certainly make history and will be a real break-through."[37] Apparently, the surgery was successful (insofar as Mr. Wilson got what he came for), though little is known of how he felt about his surgery, if he suffered any long-term repercussions, or even how or when he died, save that he went on to marry a musician by the name of Warren Combs.[38]

The next year in 1966, Money and his team established the Johns Hopkins Gender Identity Clinic, funded by a wealthy transgender woman who called herself Reed Erickson.[39] Despite the surgery on Avon the year before, sex change operations remained almost unheard of at that point and largely rejected in the medical community and society at large. That Johns Hopkins agreed to formally house the Gender Identity Clinic is attributed to Money's efforts. According to Benjamin, Money "was probably more responsible than any other individual for the decision that such an august institution as Johns Hopkins Hospital would... endorse sex-altering surgery."[40]

The clinic was immediately very popular. The doctors received nearly 2,000 requests for surgeries over the next two-and-a-half years. They were obviously unprepared or unwilling to conduct that many operations; they performed twenty-four over that time.[41] Interestingly, of the nearly 2,000 requests, only one-fifth were from females who wanted to attempt to have the genitals of males, while the rest were all males who wanted to be females, suggesting that without the influence of mass social pressure, there are far more men wanting to become women than the other way around.

It took only a few months after the gender clinic was formally established for word to get out. By October 1966, the *New York Daily News* had a piece on Avon, writing that he "underwent a sex change operation at, of all places, Johns Hopkins Hospital in Baltimore." The hospital confirmed what had been done, and all of a sudden the press smelled a big story.[42] That's when the doctors at Johns Hopkins decided to play offense and turn the news cycle to their advantage. They contacted their friend, Thomas Buckley, at the *New York Times*, and in seeming coordination they published a press release the same day Thomas Buckley published his report. After noting how Johns Hopkins is "one of the most eminent teaching and research institutions in the country," Buckley spoke of how John Hopkins was the first American hospital to grant "official support" to sex change surgery.

Soon enough, there were headlines across the nation pushing the same line about how the reputable Johns Hopkins Hospital approved of the procedure. Instead of controversy, the media promoted more acceptance of sex change surgeries, and the doctors remarked that their plan had succeeded "exactly as hoped." "The prestige of the *New York Times*... set the tone for all the other papers," they said.[43]

The impact of this positive press reception was felt almost immediately. A new gender clinic was formed at the University of Minnesota Medical School, which also worked closely with the press after the positive coverage Johns Hopkins received. Soon, there were more programs at Northwestern University Medical School, Stanford University, and University of Washington in Seattle. In a little over a decade, doctors at American universities had conducted more than 1,000 sex change operations, and there were roughly twenty centers for sex change surgery in the United States.[44]

Ironically, thirteen years later Johns Hopkins shuttered the clinic after one of their own studies failed to find evidence that sex change operations benefited transsexual people.[45] But by that time, sex change surgeries were becoming much more commonly practiced. Johns Hopkins had already granted sex change surgery its weighty imprimatur. There was no going back.

Dr. Bowers, the transgender surgeon I spoke with at his clinic in California, can trace his own academic lineage back to John Money and Johns Hopkins. His mentor, Dr. Stanley Biber, performed his first sex change surgery in 1969, using hand-drawn notes from the doctors at Johns Hopkins as a guide.[46] "He had worked on a local social worker when she had asked him to do a surgery," Dr. Bowers told me. "He at the time had no idea what that was, but, listened. He was very empathetic and was one of these people that really understood the plight that this person had."

At the time Dr. Biber was in Trinidad, Colorado, a small city north of New Mexico that, thanks to Dr. Stanley Biber's work, would become known as the "Sex Change Capital of the World," especially after the Johns Hopkins clinic closed. Dr. Bowers criticized the study that led to the shuttering of the Johns Hopkins Clinic as something done with "a political agenda to end the program" based on some sort of "fundamentalist kind of Christianity angle towards it." I found it odd that Dr. Bowers would so quickly attack Christianity while she never once criticized, or even *mentioned,* John Money's support for pedophilia. I guess some things are worthy of condemnation and others aren't.

THE REIMER TWINS

As for John Money, he had a long career before his gender clinic closed. And I soon learned I hadn't even heard the most terrifying part.

"The real case that came to Dr. Money that really made his career was the case of the Reimer twins, who you may have heard about. Quite famous. Not famous enough, though," Dr. Grossman said. The Reimer case "was the case of a family that had twin boys. Perfectly normal twin boys, but when the twins were eight months old and they went to be circumcised, the first twin whose name was Bruce, something went wrong with the machinery, the equipment, and his penis was burnt off. Essentially, he was left without a penis." I paused, looked down for a brief moment, and shuddered at the very thought.

Dr. Grossman went on. "They stopped and didn't do a second circumcision on the other twin, as you might imagine, and the parents, of course, didn't know what to do. How are they going to raise this child? So some months later, they heard about Dr. Money. Money was telling the world about his theory that a boy could be raised as a girl and do just fine, and vice versa... In the debate of biology, nature versus nurture, the environment that the child grows up in, it's all nurture. So he was convinced that even someone that doesn't have ambiguous genitalia, even someone with normal genitalia, normal chromosomes, is still born like a blank slate and could go either way, male or female."

I could imagine the Reimer family feeling immense relief when they reportedly saw Money on TV discussing sex change surgeries at Johns Hopkins.[47] Money was on a sensational program called *This Hour Has Seven Days* discussing his surgeries. They even brought in a post-op transsexual who testified that before getting surgery in an attempt to make himself a male, he "was never complete."[48] Money went on in the program to call intersex individuals and people who hadn't yet had surgery to make their genitals match their psychology "unfinished."[49] Of course, it was the surgeon's job—really Money's job—to "complete" those people. Altogether, it was yet more evidence of how the media facilitated the rise of sex change operations—and how much of what Money did was driven by hubris.

That Money was such a well-respected scientist at Johns Hopkins relieved many of the Reimers' fears. In a way, Money sold them the perfect lie for their situation—a lie he probably believed, but a lie nonetheless. If biology and chromosomes and genitals have absolutely nothing to do with your gender—if, as the *New York Times* wrote "you tell a boy he is a girl, and raise him as woman, he will want to do feminine things"—then the Reimer family's problem was solved. Bruce didn't need to be a male at all. He could be raised as a girl, and he would apparently never know the difference.

The Reimers lived in Canada, so in 1967—ten months after the circumcision accident and roughly a year and a half after the twins were born—they headed down to Baltimore and Johns Hopkins, went to Money, and had him examine Bruce.[50] "[Money] told the parents, 'no problem.' Take Bruce. You will remove his testicles. You will castrate him. You will rename him and give him a girl's name," Dr. Grossman continued. "You will put pink dresses on him and give him dresses, raise him as a girl, and never, ever tell him, never tell him that he was born male."

"It was the perfect case. Perfect," Dr. Grossman said. "Where else would he find such a perfect case? Because he had two boys, little boys. They were like one and a half or two. They were identical twins, same chromosomes, same intrauterine environment and being raised by the same parents. Same nature and same nurture. Except that one of them was going to be raised (because Dr. Money would tell the parents to do this) as a girl."[51]

According to Money, surgery had to be done quickly. Gender may "not have an innate, instinctive basis," according to Money, but he conjectured that gender was only malleable until something he later called the "gender identity gate," which happened around three or four years old when he believed gender identity was more set in stone.[52] "The immature brain has greater plasticity," as he wrote in 1962.[53] So until then, scientists could essentially do what they wanted. By all accounts, Money really believed what he was saying. He didn't seem to think his radical ideas were theories. He was convinced they were true—and was willing to stake the life of a young boy on it. As such, he called the attempt to surgically change Bruce's sex "the most humane" course of action they could take.[54]

So at only seventeen months old, young Bruce went under the knife. He was castrated. The doctors formed the approximation of a vagina on him, and his parents changed his name to "Brenda." According to the top experts of the time, Bruce was now a woman.

Yet in line with prevailing gender theories, the surgery was only half the battle. The operation attempted to alter his sex. Now his upbringing—coupled with ample dosing of estrogen during adolescence[55]—was designed to change his gender.[56] Brenda, along with his twin brother (who acted as the control in this appalling experiment), went back to Money every year for monitoring up until the age of nine. They went sporadically for another few years until 1978 when Brenda reportedly "fled in panic" during his session with Money "totally unable to respond to any talk pertaining to sex or sex education."[57]

"[Money] would see [the Reimer family] and ask questions and every time he reported his findings using false names in the literature, in the psychological literature," Dr. Grossman told me. "He reported up to the age of ten that this was a complete success. And he reported that I think he called Brenda 'Joan.' He said that Joan was completely comfortable with being a female, feminine in every way, played with dolls, loved her dresses and so on and so forth." Ultimately, Money wrote that Brenda was growing up "tomboyish" but "feminine."

Other accounts reveal that Money "consistently asserted" that Brenda's sex change and adoption of a new gender identity were both complete successes, providing solid proof for his theories. By the late 1970s, sociologists and psychologists were repeating Money's claims as further proof of his theories,[58] despite the fact that even at that time there were elementary errors in what Money chose to publish, such as repeatedly failing to get basic information correct, like Brenda's age.[59]

Yet around 1978, right when the Reimers decided not to go back to Baltimore, Money stopped making much of any comments about the Reimer case. He kept his near silence for decades.[60]

What happened during those years—both when Money was seeing the twins and long after he stopped talking about them? It turns out that the façade was impossible to keep up. Brenda wasn't a woman. Only a year after the Reimers' last visit with Money, a documentary team reported that Brenda was not at all happy as a girl. The family still hadn't told Brenda the truth at that point, but they revealed behind the scenes that they were not confident that the imposition of a false gender identity on their little boy would work.[61]

Brenda's living contradiction of nascent gender theory would be laughable if what was done to him wasn't so cruel. He apparently wanted to be a mechanic and had a "very masculine gait" according to the documentarians.[62] Throughout his childhood, he never felt like he was the gender all the authority figures in his life told him he was.[63]

When Brenda was finally told the truth and old enough to speak for himself, he rejected the lie he was forced to live. He switched back to male, hoping to recover his true identity. He began taking testosterone to counter the doses of estrogen he had been forced to receive for so many years. He had the breasts un-naturally imposed on him removed. He had a phalloplasty in an attempt to regain the penis he had lost. Eventually he got married to a woman and adopted his wife's children.[64]

Most symbolically, he changed his name—not back to his given name of "Bruce" but to "David." "He picked that name, he said, because it symbolized that what he went through all those years was like David fighting Goliath," Dr. Grossman said. "Every day, fighting who he felt he was, having to play with dolls, he didn't want to play with dolls. Dr. Money had said everything was so successful; no, he was stealing his brother's trucks to play with. He didn't want to wear skirts. He would even pee standing up."

David wasn't the only child Money abused. In 2000, after the truth of David's case had gone public, another victim named Kiira Triea wrote in strong defiance, "As one of John Money's former intersexed 'non-human' experimentees, I speak from personal experience to assert that the outcomes of highly anomalistic cases of children who are deemed available for experimentation, for whatever reason, actually provide no useful data. Unless of course there are still some holdouts who require more evidence of the strength of the human spirit."[65]

Another person self-described as a "hermaphrodite with am-biguous genitalia" wrote that she was "glad that Rosenthal [a writer who wrote on the subject of Money] is so forgiving of Dr. Money. I, however, and the thousands of other intersexuals who have had their lives so adversely affected by him are not so forgiving."[66]

As for David Reimer, he soon revealed that castration and forced femininity were far from the only crimes Money committed against him. Phil Gaetano, writing for Arizona State University

School of Life Sciences, recounted the terrifying details. "During the twins' psychiatric visits with Money, and as part of his research, Reimer and his twin brother were directed to inspect one another's genitals and engage in behavior resembling sexual intercourse," Gaetano wrote. "Reimer claimed that much of Money's treatment involved the forced reenactment of sexual positions and motions with his brother."

Gaetano uncovered even more unsettling, lurid details. "In some exercises, the brothers rehearsed missionary positions with thrusting motions, which Money justified as the rehearsal of healthy childhood sexual exploration." Reimer said that at least at one time, Money photographed him and his brother as he forced them to engage in these practices. Meanwhile, Money and up to six of his colleagues would observe.

"Reimer recounted anger and verbal abuse from Money if he or his brother resisted orders, in contrast to the calm and scientific demeanor Money presented to their parents," Gaetano wrote. All this was done when the children were in preschool and elementary school.[67]

DEFENDING THE INDEFENSIBLE

This wasn't science. This sounded more like a man satisfying his sexual fantasies. The Nazi madman Josef Mengele entered my head, and I couldn't help but feel that what Money did was a sort of anesthetized, less murderous, sexual extension of the Mengele twin experiments. The connection was even more tangible than I thought. Almost immediately after the Johns Hopkins Gender Identity Clinic founded by Money was shuttered, one of Money's protégés, Dr. Fred Berlin, began a sexual disorders clinic at Johns Hopkins—a clinic explicitly recognized as an extension of

Money's work.[68] In Dr. Berlin's office at Johns Hopkins, there was a photo of a baby Adolf Hitler.

In an interview about it, Berlin claimed he put it up "not because I have respect for the horrors of Hitler," but rather because saying Hitler is evil doesn't advance understanding. The photo instead prompts him to ask, "what about his [Hitler's] life experiences? What about his biology? Was there some sort of psychosis?" I wasn't sure how much the questions mattered, but knowing that photo was there sat poorly with me either way.

I went back to the pro-trans Dr. Forcier. Earlier she told me John Money was "hugely powerful" through his early gender work. How could she have not mentioned the Reimer twins? The moment I brought it up, she didn't deny knowledge. Of course she knew. This was her field. But she did quickly change her tune.

"I don't find that John Money [is] other than a historical lesson and a cautionary tale of how medical persons can do harm—how medicine, like any other aspect of the world, can misuse or use information for selfish or other purposes," Dr. Forcier told me. The problem wasn't in the root ideas that led to this evil. The problem, according to Dr. Forcier, was merely that "physicians have no business making decisions about gender identity until patients are ready to tell us what their gender identity is."

Dr. Forcier wasn't the only one to treat this sadistic psychopath with kid gloves. Money himself spent years making everyone believe the Reimer case was a complete success and never faced up to what he did. When he finally was confronted about it, he claimed that criticisms were based on anti-feminist and anti-trans bias.[69] "There was never any retraction," Dr. Grossman informed me. "Did he have the integrity to step forward—even though this was decades later and his ideas about gender had already permeated psychology and society? ... He never stood up, like a man, and said, 'Listen everybody, this wasn't as I said it was. Joan was not happy. She [sic] was miserable. And therefore... nature is important. It's not all nurture."

Others made excuses or ignored the Reimer case entirely, all in a seeming attempt to defend the trans movement and to protect the reputation of the father of gender theory. The authors of the biggest resource I used to learn about Money—Lisa Downing, Iain Morlan, and Nikki Sullivan (the title of the book itself is too crass to print)—merely said that, "Money's career was also beset by ethical controversies, exemplified by the internationally publicized case of David Reimer."

"Ethical controversies"? That's it? They went on to equivocate further: "Reimer's story was held variously to show Money as humane and barbaric, naïve and deceitful."[70] I struggled to see what in that entire story could possibly be deemed "humane."

Dr. Anke Ehrhardt, a tenured professor of medical psychology at Columbia University, wrote a year after Money's death that Money was "a leader" and a "truly original thinker" who received "sixty-five world-wide honors, awards, lectureships, and degrees." Without ever mentioning the Reimer twins, Dr. Ehrhardt wrote that criticisms and rejection of Money and his work were mere "expressions of hate against him as a person" even though he was "unique in his understanding, expertise and knowledge, and tolerance and counseling."[71]

Salon, a bastion of feminism and the trans ideology, went further, writing that "To Money's defenders and friends... all such criticism are liable to be interpreted as misrepresentations of Money's noble project, or attempts to scapegoat a brave pioneer whose ideas could be 'simply too intellectually demanding to pursue.'"[72] *Salon* mimicked Money's own defense, condemning critics of Money for referring to Money in an "othering" way—a common left-wing rhetorical defense to suppress criticism.[73]

These defenses worked. After some initial aversion to Money and his work when the Reimer case first came to light over two decades ago, few have ever heard of what Money did. In fact, as Dr. Ehrhardt put it, "the pendulum has already started to swing back to give John Money the proper credit for his extraordinary contribution to the field of...sex research."[74] Dr. Grossman put it in even grander terms: John Money has "made an impact on the world beyond his wildest dreams."

I set out to learn the truth about gender theory as part of my larger project to determine what a woman is. That led me to look to the history. I realize now that the reason people don't talk about the history of gender theory is not because the information isn't available, but because the details are shameful, immoral, disgusting, and—to use a term gender theorists hate—perverse.

So far, I learned that gender theory is the brainchild of sick people like Alfred Kinsey and John Money, people whose experiments masqueraded as science and who used their positions of authority to overthrow the sexual status quo and normalize what good people know instinctively is wrong. When they were done, they had left the door open to everything from sex change surgeries to pedophilia to the open sexualization of children.

Somehow their ideas began moving from subversive to tolerated to accepted to mainstream. To find out how that happened, I turned to the left's most potent instrument: the education system.

HOW GENDER THEORY BECAME THE CORE CURRICULUM

INSANE LEFTWING IDEAS have this strange tendency to move from fringe academic theories into mainstream society. It has happened so often—and sometimes with such great speed—that it can almost seem inevitable. Yet the advance only seems like a blitzkrieg because regular Americans were too busy living happy, well-adjusted, normal lives to notice what was lurking in the shadows. Now, by the shadows, what I am actually referring to is one of the most nefarious institutions in human existence, something like a fallen angel that was once beautiful and served a high and noble purpose but is now devoted almost entirely to the corruption of the youth and the deconstruction of civilization. Of course, I am talking about the universities.

THE GENDER THEORY INFILTRATION OF HIGHER EDUCATION

Take almost any terrible modern idea from communism to critical race theory, and you can clearly trace the roots back to the universities. These once-proud institutions that trained the next generation to be principled, profound, and wise leaders of men are now little more than decadent, cash-infused purveyors of prurience, vice, victimology, and a distinctive hatred for the things that give life meaning and happiness—things like family, place, religion, tradition, or even the fundamental virtue of self-control.

It was through the university that gender theory began to take hold of the American mind. Yet while it feels like the dam preventing gender theory from washing over every aspect of our culture only broke in the last few years, the truth is that there were cracks appearing all over the place for decades.

We've already learned how Alfred Kinsey and John Money began to popularize their ideas of sexualized and genderless children as far back as the 1940s, 50s, and 60s from their respective perches at Indiana University in Bloomington and Johns Hopkins. Over the next few decades their ideas metastasized throughout academia, adopted and expanded upon by popular intellectuals and unremarkable academics alike.

One of the most notable was Judith Butler, a graduate of Yale University and a philosopher at UC Berkeley. Born of Jewish parents in Cleveland, Ohio, Butler (who apparently prefers the pronouns "she/they,"[1] which makes absolutely no grammatical sense) became an extremely influential personality in the burgeoning field of gender theory over the past three decades.

In 1990, she published a book entitled *Gender Trouble: Feminism and the Subversion of Identity*. Ultimately, her influential argument was an extension of the early Money thesis that gender, as distinct from sex, has no standing in nature. Instead, gender is merely "performative." Whereas Money argued that gender is not innate and gets locked in at an early age as a result of upbringing (and, as he would later posit, hormones), Butler went one step further. She took the heart of Money's thesis but extended the timeline and proposed that the locus of gender changes was not only the outside influences of nurture but also a personal process of action and behavior. Gender is the result of how we talk, act, gesture, dress, and behave.

As Butler put it, "to say that gender is performative is a little different because for something to be performative means that it produces a series of effects. We act and walk and speak and talk in ways that consolidate an impression of being a man or a woman."[2] Or to put it another way, she said, "We act as if that being of a man or that being of a woman is actually an internal reality or something that is simply true about us, a fact about us. But actually it's a phenomenon that is being produced all the time and reproduced all the time. So to say gender is performative is to say that nobody really is a gender from the start."[3]

If gender is performative, then obviously gender is something much more fluid and not set in stone at three or four years old, as Money hypothesized. Instead, gender is the result of a continual dialogue between the person and his or her surroundings to create a masculine or feminine identity, whatever that means now.

As a result, Butler believed that "gender identity" doesn't really exist—at least not as something we are born or created with. It is something we create ourselves. She wrote, "There is no gender identity behind the expressions of gender ... identity is performatively constituted by the very 'expressions' that are said to be its results."[4]

When you couple this idea with the proposition that what we think of as masculine and feminine are both social constructs, this would seem to mean, at least to pro-trans ideologues, that gender is just as meaningless a term as "sex." Gender is our acting in certain categories that are completely arbitrary, the logic goes, which I guess would make gender itself arbitrary. All the world's a stage, and all the men and women merely socially-constructed gender-fluid players.

While Butler and Money obviously disagree, Butler's theorizing would not have been possible if Money had not already so thoroughly unmoored the idea of gender from the reality of biological sex. As Dr. Milton Diamon, a professor of reproductive biology at the University of Hawaii said, "Judith Butler and others were all very supportive of John Money, because he was saying what they wanted to hear."[5]

Now, I would quote Butler more thoroughly in order to have her make her points for herself, but the truth is I may have exhausted everything intelligible that she's said on the subject of gender. Ninety-nine percent of what Butler wrote is impossible to understand. Here are some samples so you know I'm not messing with you.

"Gender is not to culture as sex is to nature; gender is also the discursive/cultural means by which 'sexed nature' or 'a natural sex' is produced and established as 'prediscursive,' prior to culture, a politically neutral surface on which culture acts,"[6] Butler wrote.

Or there's this gem, which Butler wrote in 1997: "We do things with language, produce effects with language, and we do things to language, but language is also the thing that we do. Language is a name for our doing: both "what" we do (the name for the action that we characteristically perform) and that which we affect, the act and its consequences."[7]

A good rule of life is that the less clear someone is when they speak, the more they are trying to hide. My theory is that Butler's lack of clarity helps keep her opponents—those unwilling to spend hours trying to understand her—from being able to interpret what she is saying. That means only true devotees will be able to read and spread her theories, reducing popular backlash while allowing her ideas to spread among the most "well-educated" largely unhindered.

While this was my first time actually looking into Butler's theories, the bones of it did sound awfully familiar. I had already heard iterations of it in those I had interviewed, most notably in Dr. Forcier, the professor of pediatrics up in Providence, Rhode Island.

In fact, Dr. Forcier had repeated almost a summation of Butler's theory of gender performativity when I spoke with her. "Gender is about who you are... and there are so many different aspects of our identity there," she said. "Who you are in your family, who you are with your peers, who you are at school or work, who are you in terms of your religious and spiritual views? Who are you in terms of your gender? How masculine, how feminine? How neither? How both? How different non-binary you might feel as well as sexuality, in terms of who you're attracted to and who you want to have sex with." Everything she said was relational and about how we act in the world.

Since gender is about your interactions with and relationships to others, Dr. Forcier described how she would walk kids and patients through a transition process. "What would it mean for your kid to have short hair or to wear more masculine-type clothing or, you know, if your child wants you to call them [sic] 'Sam,' how would that feel?" Dr. Forcier said as if she were talking to the parents of a supposedly transgender child. "And for a kid," she continued, "they might be able to say that 'I'd feel great. I would feel I really like that.' And a parent might say, 'Well, let's try that and we can do that at home. Do we want to do it at school or do we want to do it with relatives? There're all kinds of decision points about how to share different parts of this journey." The "journey" Dr. Forcier describes is directly related to how the child performs to the outside world. It's Judith Butler in action.

Dr. Forcier extended the reach of gender even further, saying that, "Gender happens before kids are even actually born, and gender progresses." But in her view, it's also a process that never ends. "I take care of patients in their elder years, in their 70s and 80s, who are making decisions about their gender identity," she told me. "So, of course, gender is developmental, and of course, everybody's developmental pathway is going to be different."

GENDER THROUGH THE EYES OF THERAPY

It dawned on me that maybe I needed to expand my search. I was talking to all sorts of doctors thinking maybe the answer to the question "what is a woman?" was in science. I turned to history thinking maybe I would finally find clarity if I could find out why there was so much confusion about womanhood today. I had learned a lot so far, but obviously I still couldn't pin down from anyone what exactly a woman is. If anything, I was only more confused as different people told me that being a woman was essentially nothing. It was a performance. It was a feeling. It was relational.

Then I had a bright idea. Perhaps all along I was asking the wrong people. If Butler is right and gender really is just performative and relational, I needed to talk to a therapist—someone who is truly an expert in relationships and helping people wrestle with their identity.

That's why I sought out Gert Comfrey, a marriage and family therapist in Nashville, Tennessee, and probably one of the nicest crazy people I have ever met. Comfrey offers what is known as gender affirmation therapy from the inviting place of her own home—a home filled with the sorts of environmental justice posters and vaguely meaningless positive energy quotes chock-full of progressive buzzwords that you'd expect from a true woman of the left. One particular favorite: "deep within our roots during this time of transitional power, we seek equitable distribution and healing in this world. We invite the steady flow of sex, pride, self-power, and passion so that energy can move freely to where the collective needs it most."[8]

You'd almost think that this means something until you take a second to think about it. That's also true of almost everything Gert said during our conversation.

Comfrey told me with an incredibly inviting smile that gender is really a process of self-exploration. "So for me, like, truth, or like, reality is, like, pretty relative. So, like, my truth could be very different than your truth, or than, like, someone else's truth," Comfrey said. What does that have to do with gender? She continued. "I guess it goes back to, like, trusting the person to tell me, like, ok, this is who you know yourself to be."

There is no "this is who you are," "this is how I was born," or "this is how God made me." To Comfrey, gender has its roots in self-definition.

For Comfrey as a therapist, it all boils down to a process of listening and asking questions to help people explore if their gender identity really isn't as stable as they may have thought. Maybe people have been defining themselves too narrowly.

"So in my working with people, there's this grieving process of like, oh, gosh, like, I wasn't afforded gender expansive of play when I was a child," Comfrey told me. "Maybe I would have

71

known or felt a more connection to, like, the gender that I know now back then." This seemed to confirm Butler's ideas that gender is something performative—something you experience and interplays with your surrounding and actions.

Comfrey added that if a biological male (my words, of course) comes up to her saying, "I'm, like, a trans girl," that leads Comfrey to start asking questions. "Ok, what does that mean to you? What does that mean to you if you think you are a girl? What does that mean in terms of, like, how you want to show up in the world... Does that inform the clothes that you want to wear? How do you want to present in the world?"

I see. So, nowadays gender isn't something innate or learned. If Dr. Forcier and Comfrey are right, then it's all about how we present ourselves and how our idea of gender develops over time. Gender is performative.

But this didn't match what I was hearing out in the world. We are told repeatedly that a so-called trans woman (also known as a biological male) is a "woman trapped in a man's body." We need to embrace and even celebrate trans people because they can't help it. They were "born this way."

Dr. Bowers gave me a version of that when I asked him where exactly the line is drawn between a feminine boy or a boy who is transgender. He quickly responded that, "it's clearer than you think, at least in the mind of a child." That certainly doesn't sound like something developmental or performative. It sounds rather definitive—like the child realizes something true about himself that exists in reality, not simply in his own self-definition. I asked Dr. Bowers about his own transition, and he confirmed as well that, "I had a clear feeling, as you know, as my earliest memories, really. So early childhood."

So is gender a fluid, lifelong journey? Or are people born the wrong gender? Is it even possible to believe both at the same time? Nobody I spoke with seemed to notice the apparent contradiction.

I began to see that the gender theory movement isn't some monolithic force. It's not a pure formula with set processes and answers, like Marxism, algebra, or the guarantee that if you throw your socks in the washing machine at least one will get lost. Consistency matters much less to gender theory than that sexual norms and traditional modes of masculinity and femininity are disrupted. In the end, it doesn't matter why someone says he or she is transgender, only that you accept it. Almost everyone I had spoken with was a relativist. But as I later learned, their relativism stopped where contradicting their opinions began.

But back to the story. I still needed to know how gender got into the school system. It was clear to me that Butler's opaque philosophizing had really infiltrated the minds of the highly educated people I was talking to, whether they knew it or not. Somehow gender theory percolated everywhere, and it wasn't hard to find out how that happened.

Since 1990, the number of people awarded women's and gender studies degrees from American colleges increased by more than 300 percent. In 2015 alone, more than 2,000 students graduated with women's or gender studies degrees.[9] Contrary to popular belief, gender studies majors do actually get jobs. Sure, they don't work in a factory or in retail or in telecommunications. They aren't exactly qualified to be bankers, lawyers, insurance actuaries, customer service representatives at a call center, or ditch diggers. But they do become the heads of human resources departments. They become professional community organizers and activists. I'm sure a few become therapists like Gert Comfrey.

And a lot become teachers. Indeed, the online career finding service made teaching the number one job they recommended for graduates with a women's and gender studies degree.[10]

But it's not like gender theory appeared out of thin air in classrooms across America. Teachers didn't just all of a sudden sit down on their own and draft new curricula on gender exploration for kindergartners. School boards didn't have a vote and decide to replace honors biology with "Intro to Judith Butler." Yet somehow gender theory is now everywhere in our kids' schools. How did it get there? It came through the well-worn avenues of sex education.

THE PERVASIVENESS OF SEX ED

It may be hard to believe, but there was a time when society didn't think that schools were the proper place and young children were the proper age for sex education. That changed in the middle of the 20th century, and it can be traced back in many ways to Alfred Kinsey.

"Behind the whole sex education movement was Kinsey ideology," said Dr. Grossman, who has actually devoted much of her research to sex education. The ideology was "that kids are sexual and that, therefore, kids need to know about sex and about masturbation and about all these other activities." As one historian put it, "the Kinsey research set the tone for open investigation and instruction about a full range of sexual expression."[11]

Kinsey and Money and the other sex researchers at that time believed that nothing sexual was off the table. Moral codes were prudish anachronisms. Limitations on sexual expression were harmful to health. "Our sexual behavior... is like other animals," wrote the President of Sexuality Information and Education Council of the United States (SIECUS), an extremely influential organization promoting sex education in schools. "There is essentially nothing that humans do sexually that is abnormal."[12]

As such, Kinsey and Money's work led to the introduction of sexually explicit materials into the classrooms of universities. Dr. Judith Riesman, the researcher who uncovered Kinsey's sexual abuse of children, wrote that "the introduction of pornographic films into medical training, and the unwholesome influence of the films on individual doctors and the profession as a whole, were brought about by Kinsey."[13]

Dr. Vernon Mark, a professor at Harvard Medical School, traced the connection further, connecting Kinsey and Money to pornographic material in schools: "Kinsey seems to have provided the impetus for showing sex movies to medical students, and in 1967 they got to look at the materials from the archives of the Institute for Sex Research. Soon after, Professor John Money compiled an illustrated presentation called "Pornography in the Home," which became very popular with students at Johns Hopkins Medical School. Since Johns Hopkins enjoys a leadership role among American medical colleges, it is not surprising that roughly 90 percent of medical schools followed its lead in initiating explicitly sexual films as part of the curriculum for their students."[14] Much like Money leveraged the prestige of Johns Hopkins to normalize sex change surgeries, he used the same method to push pornography into education.

But Kinsey and Money's theories could not be kept to college students alone. Remember, Kinsey believed that children were sexual even as infants, and that sex should be governed not by any moral law, but by the dictates of personal freedom. Children weren't innocent beings meant to be protected, but rather budding sexual entities that must be awakened to their sexual nature.

SIECUS was the entity that most aggressively turned Kinsey's ideas about sex education into a reality in kindergarten through twelfth grade education—and SIECUS's roots were, let's just say, seedy. "[Kinsey] was pals with the individuals that were at the helm of creating the sex industry," Dr. Grossman explained to me. "Hugh Hefner, for example, gave money toward creating an organization called SIECUS."

Fornicate early, fornicate often, fornicate in every way possible, was the Kinsey and Hefner mantra. And the best way was to start young. SIECUS was the instrument to do that.[15]

"They say that they're teaching it so that kids will have a healthy feeling about their bodies and not be ashamed," Dr. Grossman explained. "But really what they're doing is they're breaking down the inborn sense of shame that children are born with that makes them want to cover up in front of a stranger. They want to break that down. Because if you do that, it's then easier to present more and more and more material."

For SIECUS, that material is purported to demonstrate a value-neutral, scientific way of approaching sex education, which Dr. Grossman contends is a farce. "Sexuality education is laden with values. You can't separate... If you say that you have a sex education curriculum with no values, well, then that is the value that there's no values."

Regardless, SIECUS advances this ploy through the production of educational guidelines, as they call them, which offer recommendations for how schools can offer "comprehensive sexuality education" for students form kindergarten to twelfth grade. Notice it's not just "sexual" education; it's "sexuality" education. "Sex" could be limited to mean the understanding of reproductive organs and sexual health. "Sexuality" on the other hand, is much, much broader.

These guidelines include calls for sections on such subjects as contraception, abortion, sexual fantasies, and masturbation—all in an attempt to help cultivate in students a radically open and permissive understanding of sexuality. The only reason anyone would conceive of this as being acceptable for young children is thanks to Kinsey's "research" on pre-adolescent orgasm and his proposition that people are sexual from birth.

The guidelines assure students that "being sexual with another person does not mean that masturbation must or should stop" while also making sure that students know that "young people can buy nonprescription contraceptives in a pharmacy, grocery store, market, or convenience store," and that "in most states, young people can get prescriptions for contraception without their parents' permission."[16]

Pretty much all the statements in the guidelines sound factual. But it's obvious that they are coming from a specific perspective. Yes, it's true that nonprescription contraceptives can be bought in a grocery store, and that you can get prescriptions for contraception without a parent's permission. But why does a fourteen-year-old need to know that? Why would an eight-year-old? The obvious implication is that there's nothing at all wrong with a fourteen-year-old having sex—and that there are ways to manage sex and make it "safe."

The argument is that young children need to know this because they are naturally sexual anyway and repressing them could do them damage. "In other words," Dr. Grossman said, "we would be harming [children] if we stood up and said, no, these are innocent children; they don't need to know this until they are older."

While the SIECUS guidelines do flaccidly attempt to portray themselves as neutral by adding statements like "people have a variety of beliefs about the ethics and morality of abortion" or "some religions teach that sexual intercourse should only occur in marriage," they rarely, if at all, discuss the profound spiritual, moral, health, and emotional harm that people can cause themselves even when engaging in purportedly "safe" sex. (As a quick addendum, immediately after their statement that some religions only allow intercourse in marriage, SIECUS quickly adds that "there are many ways to give and receive sexual pleasure without having intercourse"—an obvious attempt to break down chastity by degrees, if it's not possible to destroy it altogether immediately.)[17]

According to Dr. Grossman—who spent twelve years on the staff of UCLA's student counseling service—sex ed really isn't about health at all. It's all part of a full-fledged pro-sex agenda. And she saw the harm in the students she treated. "It's not about staying out of the doctor's office or else why would all these kids be like filling my office at UCLA with their STDs and their abortions? Because their sex education was about sexual freedom," Dr. Grossman said. "All of that—the infection, the distress, the depression, the anxiety, the anger—that was 100 percent avoidable" if students were taught about sex properly instead of just pushed to have sex in any and all ways from as young an age as possible.

The Kinsey-infused SIECUS guidelines would cause enough damage themselves if they were the only component of sex education. But when Dr. Grossman spoke about breaking down shame in children so that you can introduce them to more and more materials, she wasn't referring to showing children more academic guidelines or having them read more sex-positive statements presented as facts. The truth is much more disturbing.

"Matt, it's come to the point where we have this book," Dr. Grossman said, reaching to her library shelf and handing me a brightly colored children's book entitled *It's Perfectly Normal: Changing Bodies, Growing Up, Sex and Health.* An array of pre-teen friendly cartoon people appeared on the cover along with the text "Over one million copies in print" and "For ages ten and up." The images on Amazon don't show much of the inside of the book. There's a reason why. "I will just direct you toward one of the pages," Dr. Grossman said.

It took me a second to process what I was looking at. The first page featured a couple of dozen full frontal, nude depictions of men, women, and children. Dr. Grossman directed me to another page. It was a side view of a man and a woman having sex. Another page depicted a naked woman from behind, bending over and using a mirror to inspect her vagina. One passage tried to put sex in playful and fun terms for the young audience of the book: "Sexual intercourse happens when two people—a female and a male or two females or two males—feel very sexy and very attracted to each other... When a female and a male are so close that the male's penis goes inside the female's vagina, the vagina stretches in a way that fits around the penis."[18]

"It's unspeakable what these people have done to our children," said Dr. Grossman as I lifted my eyes from the pornographic book in front of me. "If this is shown by a teacher to her students in fourth grade—ten years old is, I guess, fourth grade, fifth grade—it gives a message to the student that, number one, this is ok. Number two, school is where I learn about these things, not home. I learn about my body, about sexuality, about gender, about all these things at school."

(Divorcing sex education from the family environment is critical to Kinsey's theory of liberated sexuality. The less parents know, the better. As Kinsey is reported to have said frequently, "The difference between a good time and rape may hinge on whether the girl's parents were awake when she finally arrived home.")[19]

This wasn't the only pornographic material being shown to our children. Recently at the Dalton School, a private school in Manhattan, teachers showed six-year-olds—six!—a video depicting sex organs and solo sex acts, along with a young boy asking his teacher why he feels like pleasuring himself sometimes.[20] Girls in California as young as ten and eleven have been taught how to put condoms on models of an adult male penis while the boys in their class watched. They were also being taught how to engage in oral and anal sex with their "partner."[21] A couple of years ago in Virginia, parents were shocked to find out that their schools' classrooms and libraries were filled with sexually explicit books that included depictions of masturbation, oral sex, sexual intercourse, incest, rape, and underage drinking.[22]

Sexual health guidelines have served as a vehicle for this kind of filthy material to filter throughout American schools, often without parents having any idea what's going on. The SIECUS, for example, has distributed over 100,000 copies of their guidelines and another 1,000 copies are downloaded from its website every month.[23]

This isn't just a recent phenomenon. As far back as 1968, New York University received a federal grant to establish a graduate level sex ed training program for teachers.[24] By 1994, SIECUS had eighty nonprofit organizations as members all pushing comprehensive sexuality education in school—including the YWCA, the

American Medical Association, and the American Library Association.[25] Only two years later, forty-seven states required or recommended sex ed programs in their schools, and every single state required a program on HIV and AIDS.[26]

Parents and some educators tried to resist the imposition of sex education for a time. At first there was the wholesale rejection of sex education in some corners—particularly among Christian groups. Later, there was a transition to promote abstinence only or abstinence-focused sex education. Now, by all appearances, more virtuous-minded parents are happy if they just can opt their children out of the most explicit materials.

Today, sex education is so ubiquitous many people think it's completely normal, and even laudable to teach younger and younger kids about ever more novel and obscene types of sexual acts and expression. Dr. Grossman noted that "[only people who] maintain some connection to a different system of belief, a different system of values and morality—for example, a Judeo-Christian system—...[can] say that, 'no, no sexuality is not for kids, that children need to be protected from that sort of material', that their innocence is precious, and that it's a crime to take that innocence away."

HOW SEX ED LED TO GENDER ED

After comprehensive sexuality (remember, not "sexual") education was firmly established in the school systems, it became a perfect vehicle for gender ideology. All it took was gender ideology to become a standard and accepted part of leftwing social reform. Then, gender ideology entered the education system the same way that extreme sexual openness did, moving first from scientific discussion then to academic dogma and finally to a mandatory part of public education. The only difference is that it happened a lot faster.

After people like Money and Butler broke ground in the theory of gender, the basic tenets of gender ideology moved from radical theory to scientific "fact." The big shift came in 2012.

"So what is gender identity disorder? Is it a mental illness?" I asked Dr. Grossman.

"Well, we're not supposed to use that term anymore as of 2012, but prior to 2012 we used the term to describe a disorder that was very rare," she responded.

"And now that label doesn't exist anymore? We don't talk about 'gender identity disorder.'"

"Well, it depends who the 'they' is," she said. "But in general, right, the diagnosis has been rejected, and the new diagnosis is gender dysphoria."

What's the difference? Dr. Grossman explained: "The main difference between gender identity disorder and gender dysphoria is that the pathology is no longer that the child is questioning who they [sic] are. Rather, it's the distress that the child feels, which is due to society, the parents, the school, not accepting those feelings of the child." With gender identity disorder, the issue lies in the person. With gender dysphoria, the problem is everyone else.

I presumed the change was based on some new research or theoretician or something like that. The truth is, it was political. The change was made in what is called the *Diagnostic and Statistical Manual of Emotional Illness,* or DSM—a sort of bible for psychologists to diagnose their patients.

What happened in 2012 was that the DSM was updated to DSM-5—and with it "gender identity disorder" was dropped and "gender dysphoria" was in.

As a certified psychiatrist, Dr. Grossman was able to give me the inside scoop. She told me that when organizations like the American Psychiatric Association or the American Academy of Child Adolescent Psychiatry gather together, they form smaller committees or task forces of about eight to twelve people to address discreet issues—such as changing how we define gender-confused people.

"There are 38,000, I think, psychiatrists in this country," Dr. Grossman said, "who are the people that end up on these task forces, end up on these committees. Generally they're activists, especially when it comes to these issues—sexual issues and gender issues... So this is not a consensus."

Once this small group of ideologues changes the definitions, the medical establishment, professional organizations, and other psychiatrists can point to it as definitive, the final word. They use the weight of an established publication to beat back any dissent and label people as anti-science. "I'm a practicing psychiatrist; I'm a board-certified child and adolescent psychiatrist," Dr. Grossman told me. "I have tons of friends who are psychiatrists. No one asked us."

When gender confusion was redefined from being an internal pathology and into being an identity, that opened the way for teachers and the education system to more aggressively teach about fluid gender as just another part of human sexuality. Gays and lesbians had already spent decades proactively pushing to get pro-gay materials into libraries and schools.[27] The transgender movement quickly made up for lost time.

The most up-to-date SIECUS guidelines state definitively that one of the "life behaviors of a sexually healthy adult" is to "affirm one's own gender identities and respect the gender identities of others." Other sections say that "people's understanding of their gender identity may change over the course of their lifetimes" and "gender role stereotypes are harmful to both men and women."[28]

Dr. Grossman, as usual, put it more succinctly: "Kids are being taught that gender is between the ears, it's what you think, and sex is between the legs—and the two are completely different things."

The propaganda is nearly ubiquitous now. A kindergarten book in Virginia entitled *My Princess Boy* teaches five-year-olds about gender transitioning. As Anna Anderson of the American Principles Project detailed: "Other titles include, but are certainly not limited to: 'Prince and Knight' (second grade), 'Beyond Magenta: Transgender Teens Speak Out,' 'Some Girls Bind,' 'Weird Girl and What's His Name' (positively featuring statutory rape between a young boy and his boss, and a relationship between a girl and her teacher), and 'Being Jazz: My Life as a Transgender Teen.'"[29]

TV host Megyn Kelly spoke about how her son's school did a three-week experimental transgender program for third grade boys where they were told they could "take a pill to prevent puberty" and then later have their genitals "chopped off."[30] The parents didn't find out about it until after the program had ended.

Taxpayer funding is also being used to spread something called the gender unicorn. It started in 2016 in a school district in Charlotte, North Carolina, and is designed to help teachers create a "gender inclusive classroom."[31] Styled after kid-friendly cartoons, the gender unicorn thinks rainbow thoughts and between its legs is a DNA strand (as if to say that the only part of the unicorn that is determined by genetics is whatever physically happens to be in between its legs).

Demonstrating that the sick people pushing this program truly believe that no age is too early to sexualize children, the gender unicorn is designed for kids in kindergarten and even preschool. It then purports to help students understand their sexuality by showing them sliding scales of gender identity, gender expression, the sex they were assigned at birth, who they are physically and emotionally attracted to, and more. Often preschoolers can't read. They are learning things like shapes, colors, and the concept of sharing. They cry when their nervous, loving mother drops them off at school. And now they're being asked to openly explore their sexual identity with the aid of cartoon.

When the gender unicorn was introduced in schools in Oregon, the Sexual Health and School Health Specialist at the Oregon Department of Education (yes, apparently blue states have those), said, "we have more and more kindergartners coming out and identifying."[32] Apparently nobody told her that kindergartners will identify as a T-Rex or Elsa from *Frozen* if you tell them they can.

Another take on the gender unicorn is the genderbread person. Like the gender unicorn, it separates out gender identity from attraction from biological sex from sexual expression, and does so in a way meant for children.[33] Oftentimes in school districts, parents are able to opt their children out of sexual education. But the advance of gender ideology is even more totalitarian. When a parent in Oregon complained about the appearance of the genderbread person in her child's classroom and asked why her child was not able to opt-out, the principal told her that teaching on gender identity would happen in all subject areas throughout the school year.[34] So the mom retained her "right" to remove her kid from health classes, but she would have no idea when her child would be exposed to gender theory.

Something similar happened in Connecticut where a book called *Introducing Teddy* was being presented to elementary school students. As the story goes, Teddy was manufactured as a male but in his heart and mind he insists he is a female bear. Just like in Oregon, the school district told parents that they were not allowed to have control over what the state taught their children. Their kids would be read *Introducing Teddy*, and that was that.[35]

It's amazing how quickly radical gender theory went from unheard of to strange to socially accepted to educationally mandated. They push it in the schools first because they know that children are more malleable and impressionable. Not only that, but if they propagandize the children first, the parents are sure to follow.

The plan appears to be working. The classroom, after all, isn't the end for social reformers. It's the beginning. From the classroom, we've seen a sort of social contagion. You can't go anywhere, read anything, watch any movie, view any ad, or have any conversation without gender theory prodding its way in. Gender theory is the new civil religion. As I soon found out, it lays claim to everything—and heretics are unacceptable.

THE TRANSGENDER TAKEOVER

WE DON'T KNOW exactly how many transgender people there are in America. Different sources report wildly different numbers. Even the most generous estimates guesstimate that transgender persons represent a fraction of a percent of the U.S. population.[1]

You'd be forgiven for thinking there were a lot more. After all, as part of the militant embrace of diversity and inclusion, popular culture is becoming awash in transgender characters. If an alien were to turn on a random American movie, TV show, or news program, he (xe?) would be completely shocked to then be told that a majority of Americans are boring, unremarkable, white heterosexuals.

The imbalance is purposeful. The more human beings see something, the more they accept it. Sleeping around seems awfully odd if all you've observed are monogamous marriages. BDSM no longer appears strange if you've gorged on a steady diet of increasingly violent pornography. Maybe the state forcibly depriving a parent of custody over her son because she refused to allow the child to be castrated won't seem that bad if every third character on TV and social media avatar has been castrated as well—and look how happy they are!

But I'm getting ahead of myself.

FROM UNKNOWN TO CELEBRATED

Transgenderism has taken the (western) world by storm, and there's no medium where we aren't confronted with it, often on a daily basis. While this appears to have happened rather rapidly, pressure had been building for quite some time.

In the early 1950s, a former GI from the Bronx who renamed himself Christine Jorgensen went to Copenhagen, Denmark, received a sex change surgery and massive doses of estrogen, and became the first international transgender media sensation.[2] Three wire services together sent out a sum total of 50,000 words on the Jorgensen story in the first two weeks of coverage.[3] While Alfred Kinsey and John Money conducted their experiments, and the ideas behind sex change surgery and gender theory spread through academia, society at large took passing interest in the transgender issue after Jorgensen. One could be forgiven for living life in the mid 20th century without worrying if your daughter would come home and announce she is actually a man. It was a simpler time. Jorgensen was a media sensation, but like all media sensations he was soon forgotten.

In the 1970s, the controversy over Renee Richards broke onto the scene, presaging debates we are still embroiled in today. Renee Richards was a tennis player who underwent a sex change and proposed that, as a result, he was a woman. As Dr. Bowers recounted to me, "Renee Richards was a competitive tennis player who was assigned male at birth, went through a transition, and when her [sic] situation was found out, the women in the U.S. tennis circuit were very upset, thinking that she [sic] had a competitive advantage, and [they] tried to block her [sic] from competing."

Renee's case went all the way up to the New York State Supreme Court, which ruled in his favor, writing that forcing Renee to take a Barr body test to determine his sex before being eligible to compete was "grossly unfair, discriminatory and inequitable, and violative of her [sic] rights under the Human Rights Law of this state."[4]

After that, it was yet again largely quiet on the western front, until all of a sudden the trans movement stopped making splashes in the headlines and began to form a tidal wave that would wash over society. In 2008, the reality show "America's Next Top Model" featured a transgender character, normalizing transgenderism with regular appearances on many Americans TV diets. A few years later, Bruce Jenner was reborn as "Caitlyn." *Vanity Fair* ran a flattering cover story, and nearly every media outlet in America praised him as a hero. Soon after, the Obama administration lifted the ban on transgender service members in the military (a ban that President Trump reinstated and President Biden lifted yet again).[5]

Transgenderism was moving from sideshow to the main stage, and now it's in absolutely every sphere of life. There's no avoiding it. Transgenderism is the heart of the cultural conversation in America, and the only escape is death or becoming Amish.

In the movie industry, Amazon Prime produced a remake of the classic *Cinderella*, casting the fairy godmother as transgender.[6] In the *Marvel* universe—which produces the biggest blockbuster hits today—Disney (which owns Marvel) announced that there would soon be a transgender superhero.[7]

Television boasts its own array of transgender characters. The hit show *Glee* on Fox introduced a transgender character as far back as 2012.[8] A series called *Transparent* about a father of three grown children announcing he is transgender ran for five seasons until 2019.[9] *Orange is the New Black* on Netflix featured a trans character played by an actual trans actor, Laverne Cox.[10] Netflix is reportedly introducing its first non-binary (meaning not following the "binaries" of male or female) character in a series for preschool children.[11]

The celebrity world is filling up with a sizeable collection of trans or non-binary personas from Chaz Bono to the creators of the Matrix, the Wachowski brothers (now proclaimed "sisters"), to actress, singer/songwriter Demi Lovato. Lovato—a former Disney child star—is a self-anointed prophet of the sexual anarchy pioneered by Kinsey and Money. One of her Instagram posts in 2021 read, "Be a slut. Show your body. Get naked. Have all the safe, different, consensual sex you want. Be kinky. Masturbate. Make/watch porn. Make money. Just a reminder that being sexual is okay."[12]

Lest you worry that only adults and former Disney child stars are in on the fun, Desmond is Amazing is a prepubescent drag queen who was featured glowingly in a segment on Good Morning America.[13] His abuse-courting parents also allowed him to dance at a Brooklyn gay club while grown men tossed dollar bills at him. Of course, no cell phones were allowed while he danced for the men at the bar.[14]

Even if you aren't a pop culture junky, it's impossible to get by watching the news over the past few years without enduring a continual parade of transgender "firsts." Rachel Levine was confirmed as the first transgender cabinet secretary, and he now runs the Department of Health and Human Services charged with managing health policies governing, among other issues, Medicare, Obamacare, and the coronavirus. We've had our first transgender Rhodes Scholar,[15] our first transgender boy scout,[16] our first transgender national political party convention speaker,[17] and even our first high-profile transgender

2222222

government leaker and whistleblower in the person of Chelsea Manning. The amount of history being made is just staggering.

Drag Queen Story Hour swept across the country as frighteningly dressed and confused men (generally) with a strange affection for young children formed chapters and invaded local libraries across the country. Human Resource specialists and allies everywhere have taken it upon themselves to announce their preferred pronouns in emails and everyday conversations at work and in social gatherings. You can be sure if you refuse to do the same, your lack of inclusivity will be noted.

Even having a baby has become another avenue for the pervasive touch of gender theory. Hardly a hospital in America has blue or pink blankets and caps for tiny newborns anymore. They are either a mixture of pink and blue—or lack the colors at all—apparently in an effort to save that puffy-eyed nugget of a baby from having a gender imposed on her at any point in time.

There is no escape. Maybe you've already had your kids, and you're not having any more. Maybe you work from home or for yourself, and you even homeschool your kids. Maybe you've already cut the cord on the television, deleted the social media apps, and the only celebrities you can name are Elvis Presley, John Wayne, Audrey Hepburn, and Mickey Mouse. Well, that doesn't stop the parents down the street from engaging in the hot new trend of raising their children Zyler and Kadyn as genderless or genderfluid. (Zyler and Kadyn, by the way, are actual twin children in Cambridge, Massachusetts whose parents are raising them without gender so that the children can decide for themselves when they're older.)[18]

You don't just have to worry about your kids. When was the last time you had a real conversation with your spouse about gender identity? When I spoke with Dr. Bowers, I asked him how his family dealt with his transition, seeing as he was married and had kids at the time. "If you marry someone who is your best friend, they're likely to understand you, and they're likely to be with you in the long run," Dr. Bowers said. "We're still married legally," Dr. Bowers said of his wife, "but I wouldn't call it a marriage. I would call it more like a modern family."

"Is your experience typical, where the spouse is supportive?" I asked.

"Increasingly so," he said. "I think as trans as a phenomenon has emerged and become, you know, more acceptable, spouses go along for the ride." He continued, "If somebody is not a great spouse, whether male or female, and you add this into the mix, you're likely to create problems. But if you have someone who you've been friends with and you have mutual respect for, you know, it often remains." Got it. So if your spouse declares that he or she is the opposite sex and the marriage breaks down, it's really your fault for not accepting it. There were probably already problems with you in the first place.

WHAT IS TRUTH?

There was a word that Dr. Bowers said, however, that stuck out to me, if only because variations of it had appeared repeatedly with every pro-trans person I had interviewed. That word is "acceptable." A seeming key to transgenderism appeared to be this idea of "accepting" or having "acceptance" or, going one step further than that, proactively "affirming" people in their decision to change genders. In fact, it began to dawn on me that the staggeringly fast advance of the transgender movement through popular culture is driven by the twin powers of acceptance and affirmation. It all just sounds so good, so positive, that it's difficult to resist.

Nobody I spoke with wove this comforting web of affirmation more than the gender therapist Gert Comfrey. "My job is really to be a deep listener and to offer support when it's needed. How can we get creative about solutions for your life?" Comfrey told me at the very beginning of our interview. "How can we experiment with other ways of being in the world that might feel more authentic or more life-giving?"

Her words were soothing. After all, who doesn't want to be authentic or enjoy the fullness of life?

True to 1960s form, the way to be authentic and find the fullness of life, according to Comfrey, is not to fulfill your duties and obligations, rise to the challenge of human existence, or live in accordance with some ultimate truth. Rather, fulfillment comes from loosening restrictions and casting off any inhibitions that might hold the lone individual back. "How do you operate in a world where we have a pretty rigid gender binary in our dominant culture?" Comfrey posed as a hypothetical question to her clients. "Does that resonate with you? Does that fit with who you know yourself to be? Or is there something else, something beyond or in between or a little bit of both?"

Self-discovery and self-acceptance come first. This concept isn't unique to gender theory. Popular philosophies posited by everyone from Oprah Winfrey down to feel good Instagrammers promote the idea that self-actualization comes through self-understanding. Fullness comes from within. Happiness is found after first finding "my truth."

It all reminded me of that hippie song by Fame "I Sing the Body Electric." The lyrics go "I celebrate the me yet to come/I toast to my own reunion." Everything in the self, nothing outside the self, nothing against the self—modern philosophy in a nutshell.

It's all so intoxicating, so... empowering. I get to define my truth. I feel affirmed and loved and accepted. As I sat down with Comfrey... with Gert... I wondered if she could be *my* therapist. I wondered if I could open up to her like I've never opened up to anyone before. Where should I start?

"I don't mean for this to sound like a silly question, but with the fluidity of these things, no definition, how do I know if I'm a woman?" I asked, tentatively, vulnerably.

"That's a great question," she said.

I explained my musings. "I like scented candles."

"Yeah," she interjected.

"I've watched Sex in the City."

"Yeah!"

"So, how do I know?"

"Matt, that question right there, like, that question," she said, "when it's asked with a lot of curiosity, that's the beginning of a lot of people's, like, gender identity development journey."

I was already on one journey to determine what is a woman. I wasn't sure I had time to start another journey. Not only that, but the gender journey did seem fraught with peril.

Defining everything from within the self causes profound frustrations. It makes it impossible to pin down facts and ideas. I asked Dr. Forcier how we determine whether someone is male and female. She started listing a litany of everything from hormones to "how you will be read or assigned as a gender in our culture" to "how you feel." I interjected: "But in reality, in truth."

"Whose truth are we talking about?" she retorted.

"The same truth that says we're sitting in this room right now, you and I. We're not on a plane in the sky. We're not in Victorian England."

"My patients' truth isn't determined by you," Dr. Forcier said, matter of factly.

I began to understand that people like Dr. Forcier and Gert Comfrey were operating on totally different planes than I was. I was asking for truth. My questions seemed straightforward. But to them, the questions didn't even compute.

You can't combat gender theory through logic. Honestly, you can't even understand gender theory through logic. Logic means truth. It demands an external reality from which to judge actions and determine events and process experiences. Gender theory has none of that. It is relativist.

But nobody can be a complete relativist. Relativists believe there is no external truth—except for all the times they do believe in external truth. Surely they won't stick their hand in a fire, assert that "their truth" means fire isn't hot, and expect to walk away without burns. They aren't stupid. They just aren't consistent.

For no explicable reason, gender theorists have decided that gender is not rooted in reality, but other things are. The distinction is arbitrary.

"My son recently told me he's a piece of broccoli," I informed Comfrey. "I think he's been watching too much *Veggie Tales*. I'm not going to chop him up and put him in a salad, you know. So how do we differentiate between fantasy, play, delusion versus someone's actual truth?"

She skirted around the question. "I think this is where it's really important to keep in mind that, like, again, it's not just like we wake up one day and like, oh, this is who I am now. It really is a gender identity process, a developmental process that takes years," Comfrey said.

So is my son a piece of broccoli? She didn't answer, obviously not wanting to assert that truth can exist in some areas but not in others. Yet she implied that my question was erroneous, and that thinking you're broccoli is not the same as gender. They are two different types of questions. I failed to ask her what would happen if my son's broccolian identity were developed over years like how she believes gender identity develops, but it wouldn't have mattered. The assertion is that gender is different from all other types of reality. There is no explanation. There is no reason for it. That people experience discomfort believing that their gender does not match their sex is enough to reject reality's claims.

Perhaps I'm not being completely fair. After all, none of the trans activists I spoke with actually told me they believe in any truth at all. Maybe they *would* stick their hand in the fire. Maybe they *would* accept my son's identity as a broccoli. Maybe they are logically consistent.

I recently wrote a book entitled *Johnny the Walrus* about a boy who pretends to be a Walrus and his mother who believes his self-identity and tries to help him transition. The book is obviously not meant to be serious. Or maybe it's not so obvious. Perhaps I really am on the cutting edge of trans-specism, and the only reason contemporary gender ideologues don't talk about our trans-species brothers and sisters is because they aren't accepting and affirming enough.

LANGUAGE FLUIDITY

It all sounds like a joke, but it has real consequences—the most common of which centers on our debate over language.

When it comes to gender, the language is always changing. Is it "mastectomy" or "top surgery"? Is it "sex change" or "bottom surgery"? "Nursing women" are now "nursing people." "Breast-feeding" is "chestfeeding." We are told that women can have penises and men can have periods. The allowable language often changes so quickly that even transgender champions can't keep up.

"Many of the terminologies... are used now to describe people who don't necessarily fit in a completely binary system of male and female," Dr. Bowers told me.

"On your website you refer to 'male-to-female surgery', so I'm wondering..." I posited, wondering how he can say that so many are non-binary yet his surgeries are posed in binary terms.

"In all fairness, I mean, we're sometimes... I'm even behind the times. So really defining even that term is sort of past tense. So, it's sort of a nudge for us to even update our terminology." Beware: even one of the nation's busiest sex change surgeons is unable to uphold the new and ever-changing rules.

Nowhere is this terminological minefield more treacherous than when it comes to pronouns, where people simply declare that they now prefer pronouns that not only don't match their sex, but also don't even reflect the fact that they are a singular person instead of plural people. Everyone else is expected to submit to their declaration and use the preferred pronouns, even if they make no sense.

I discussed the issue with Comfrey. "So you were assigned female at birth, and you now identify... as they/them/theirs. That's your pronouns, right? What does that mean?" I asked.

She spoke of how she was called "she" and "her" her whole life and that's how she grew up. "Then I took some time with some really trusted friends to explore," she went on. "What would it be like if I heard people in my life referring to me using they/them pronouns?... And I asked people to just try that out. Can you try that out? Can you talk about me? Kind of like, can I overhear you talking about me? And the minute I heard people referring to me as 'they' or 'them', like, 'this is their house'... or 'I really like them, Gert.' 'Oh they're awesome.' Right. It really resonated... I was like, oh yeah, that feels so much better to me."

It was Butler's performative gender identity on display. Comfrey discovered dissatisfaction within herself, but she did not seek to change within herself. Instead, she asked those around her to change. Her gender identity could not be validated until the external world began recognizing her as she feels she is. The world must contort to her reality. That someone else might have a different reality—a reality perhaps that the words "they" and "them" can only refer to more than one person—doesn't matter.

"What does it feel like to be 'they'?" I continued. Feelings are always a perfect subject for nonbinary people and their allies.

"For me, it really is honoring that, like, my gender is beyond the gender binary. I'm not male or female... My gender or my gender expression is something beyond that binary," she said. "So for me it feels really expansive, really affirming. There's, like, a freedom and liberation that really shows up for me."

In her mind, she is experiencing an inversion of Jesus's words: not *the* truth will set you free, but *my* truth will set *me* free.

And who are you to deny that freedom? If her truth set her free but her truth can only be validated—can only be truly fulfilled—through the external recognition of her gender performance, then her liberation can only be achieved through your affirmation.

Some people, like me, insist on only using pronouns that match a person's sex. Others personally prefer to only use the proper, biological pronouns for other people, but they don't want to demand that others do the same. To them, if someone else wants to call herself "they" or "them" or "zir" or "your excellency," it really doesn't matter. I will call you one thing. You will call yourself another. Live and let live, is the mantra.

But that doesn't work. In a weird way, the social libertarians and the gender ideologues are operating on similar presuppositions. They both assert the primacy of freedom and self-definition. As such, it only seems natural that you are free to do what you want in your own sphere as long as you don't harm anyone else. Your freedom to swing your arms ends at my face, as the saying goes.

The disagreement comes when we ask what constitutes "harm." To the social libertarian, no harm is caused by using someone's biological pronouns instead of his or her preferred pronouns. A nonbinary person is free to use his or her pronouns; I am free to use mine. Theoretically, we should all be able to get along. But by failing to use the "right" pronouns—by failing to affirm—you subvert a trans person's identity, in his or her mind. It's not just that hearing people say "they" and "them" makes people like Comfrey feel good. Their very identity, their entire conception of themselves, demands that you use "they" and "them." Failing to use the right pronouns may not kill the body. But in their minds, it destroys a trans person's essence.

However, it is important to recognize that the interaction between transgenderism and wider society isn't a one-way street. It's not only that transgender people "discover their true selves" and then demand that others respect and affirm that self with outward displays like using preferred pronouns; society also influences gender. As Comfrey said, "A really important point [is] that [this] isn't just happening, like, in this vacuum, right. It's not just, like, ok, I have boy feelings or girl feelings. But really, like, gender is a conversation that we're having in society."

RAPID ONSET GENDER DYSPHORIA

That society and societal perceptions influence gender identity isn't just another extension of Judith Butler's theory of performative gender. It's a documented fact. And nowhere can we see this more clearly than in the shocking phenomenon of rapid onset gender dysphoria (ROGD).

A simple formula holds that when society condemns and discourages something, you get less of it; when society affirms and promotes something, you get more of it. By and large, our society is certainly affirming and promoting transgenderism, so it only makes sense that now we'll find more transgender people. But rapid onset gender dysphoria is much bigger than a natural rise in pro-trans sentiment—or even in the number of people who identify as transgender—which one might expect from increased social acceptance of transgenderism. It is a social contagion taking over entire peer groups, transforming a collection of otherwise normal girls into transgenders in one fell swoop.

Abigail Shrier, a brave and fiercely independent journalist, has explored the issue extensively. Rapid onset gender dysphoria "differs from traditional gender dysphoria, a psychological affliction that begins in early childhood and is characterized by a severe and persistent feeling that one was born the wrong sex," Shrier writes. "ROGD is a social contagion that comes on suddenly in adolescence, afflicting teens who'd never exhibited any confusion about their sex."[19]

Here's how it works. ROGD largely shows itself in young, teenage or pre-teen girls who are probably more susceptible to social pressure than any other group of people in human existence. They are already nervous, unsure about their bodies, physiologically confused, and desperate for social acceptance. That's not because they have a disease. It's called puberty and adolescence. Unfortunately for these girls, they've been convinced that what they are feeling is different from what pre-teen and teen girls have felt in every era of human existence.

To be fair, their feelings may be no different, but their circumstances are. Few other eras have been so replete with broken homes, divorced parents, unmoored social change, unstable local communities, frequent moves, lack of a common culture, faith, and experiences among peers, overindulgent acceptance of sexual experimentation among youth, and immediate access to extreme pornography, all on top of inescapable social media that allows others to pretend their lives are perfect on the one hand yet anonymously and viciously attack others in the most personal terms on the other. So while the base feelings these young girls are experiencing are common to all adolescent human existence, the social and moral environment in which current young people live seems designed to cultivate instability and unhappiness.

In these circumstances, young girls are desperate for love, acceptance, identity, and, of course, affirmation. Maybe they can't find all of that in the home. Or maybe they can, but they've been taught from a young age in pop culture and in the schools to hate and distrust their parents and anything that even faintly smells of traditionalism. That's where the idea of gender dysphoria enters in and begins to take over.

Lisa Littman, a doctor with a master's in public health, conducted research on this phenomenon, interviewing hundreds of parents whose children experienced sudden or rapid gender dysphoria that occurred during or after puberty. Dr. Littman issued a seminal report on her finding in 2018[20] where she traced how ROGD advances. "Parents describe that the onset of gender dysphoria seemed to occur in the context of belonging to a peer group where one, multiple, or even all of the friends have become gender dysphoric and transgender-identified during the same timeframe," Dr. Littman discovered. "Parents also report that their children exhibited an increase in social media/internet use prior to disclosure of a transgender identity."[21]

This is a distinctly different cohort from what used to be considered transgendered. Dr. Grossman explained the distinction. "There are individuals—rare, very rare individuals—who from an early age, from the first years of life, have consistent, continuous, severe distress from their biological sex," she told me. "These children, like I said, are very rare. They exist anywhere between one in 30,000 people and one in 110,000 people. Those are the numbers that have been established from long ago. These children suffer, and their parents suffer tremendously. They have what used to be called gender identity disorder."

Dr. Bowers confirmed that traditionally, most people who suffered gender identity disorder didn't start experiencing it as teenagers. "In the histories I take of people I ask everyone, 'When did you first feel like you were wrongly assigned?' An overwhelming majority of people say that at least before puberty they felt like something was different, and they kind of knew that something was different."

It's not just the age that differentiates those who suffer from traditional gender dysphoria and those who experience ROGD. It's also the sex. According to Dr. Littman's research, of those who said they experienced gender identity disorder later in their childhood, 82.8 percent were female, and their identity crisis appeared roughly around fifteen to sixteen years old. Yet as I previously learned, when Money opened his gender clinic at Johns Hopkins, he received 2,000 requests for gender reassignment surgery and 80 percent of those requests came from biological men—an exact inversion of ROGD. Dr. Grossman confirmed this with roughly similar numbers, telling me that "it was much more common in boys; I think five times more common in boys than girls" to suffer from gender identity disorder.

By all accounts, gender identity disorder is a real condition where people—overwhelmingly boys—truly suffer beginning at a very young age from a mismatch between their perception of themselves and the reality of their sex. "I really hope to God that you include this," Dr. Grossman urged me with slow and deliberate words. "I have the utmost compassion for people who suffer from gender dysphoria. It's a nightmare for them and their families. It's a terrible thing to suffer from."

I couldn't agree more. But she took my next thought right out of my mouth. "But just because a young person," she continued, "one day after being immersed in ideology decides that their problems, their social problems, their emotional problems, are all rooted in [that they were] assigned the wrong gender, and that all they need to do is transition and take medicine and have surgery, and then they will be a healthy human being—no. No."

Gender identity disorder does exist medically. It's not just some socially-created psychosis, afflicting adolescents. But that does not mean we should "affirm" anyone's disconnect from reality—no matter when that disconnect began.

It is also true that a condition called intersex exists, which many used to call hermaphroditism. This is where people are born with ambiguous genitalia or both male and female genitalia. Dr. Grossman informed me that intersex babies "are born in about one in 10,000 births"—which means, at a minimum, there are three times more intersex people than there are those who suffer from gender dysphoria from a young age.

These people do suffer a terrible burden where their physical anatomy may not reflect their perceptions of themselves. Often they actually do have their sex "assigned at birth" because their sex is unclear. Intersex people are an extremely small portion of the population, but their circumstances and difficulties shouldn't be ignored. Yet they are commonly elided with the transgender movement in general and associated with ideas about gender fluidity.

"Intersex people are very, very upset about how gender ideology has used some of the terms that really only apply to them like 'assigned sex' and 'intersex,'" Dr. Grossman told me. "If you follow them on social media, they're very angry because they truly are intersex, not some confused, you know, fourteen-year-old who never even heard of this a week ago and decides that that's what their [sic] issue is."

Dr. Grossman was, of course, describing rapid onset gender dysphoria, which only came into being after widespread acceptance of transgenderism and the spread of transgenderism in popular culture.

Unlike truly intersex people and those who suffer gender identity disorder from the youngest age, late entrants into transgenderism are heavily influenced by peer groups and social pressures. "Friendship cliques can set the norms for preoccupation with one's body, one's body image," Dr. Littman wrote in her study on ROGD. In the friendship groups she studied, Dr. Littman found that over one-third—36.8 percent of groups—ended up with a majority of the individuals identifying as transgender. "Parents described intense group dynamics where friend groups praised and supported people who were transgender-identified and ridiculed and maligned non-transgender people."

This intense social onslaught happens within a psychologically vulnerable population. Forty-one percent of the young people who underwent rapid onset gender dysphoria previously identified as non-heterosexual, and nearly two-thirds (62.5 percent) were diagnosed with some sort of mental health or neurodevelopmental disability disorder before they transitioned.

"You're probably aware that the kids who have gender dysphoria are most likely the ones with serious comorbidities; they have anxiety before all the gender issue comes up," Dr. Grossman said. "They're neuro-atypical, a lot of them, they're on the spectrum. They might have depression, family issues." Amid this confusion and difficulty, gender ideology gives them a place and an answer.

"They don't fit in with their peers. They've been bullied. They don't know where they belong," Dr. Grossman continued. "Then they discover a new world... on YouTube or Reddit or Discord or at school. And they discover, hey, there's this group of people and they also don't fit in. They're different, they feel different, they don't know where they quite fit in. They're not sure who they are. They're also anxious. They're also socially awkward. Gee, that's where I fit in. And there's no greater wonderful feeling, especially for an adolescent to find a peer group where you fit in."

These children are confused and troubled. Maybe they don't have good family lives. The culture at a bare minimum tells them that transgenderism is everywhere, and that it's great. If they don't embrace some sort of alternative sexuality, they are condemned. The moment they do, they are embraced. They are *affirmed.*

WHAT IDENTITY IS "ESSENTIAL" AND WHAT ISN'T?

But let's just pause for a second. Don't teenagers make radical, resolute, and often stupid proclamations about themselves and the world all the time, and we don't take them seriously? The Che Guevara commie teen still has an iPhone and Nikes. We don't actually think he wants to set up firing squads. He probably has never even held a gun. A teenage girl "literally" hates her dad and will never forgive him when he tells her she can't take his car to go on a chaperone-less weekend trip with a couple of friends, including her boyfriend. That same daughter fiercely holds her dad, comforted by him when the nimrod breaks up with her. Why do we treat self-proclamations of gender identity that came out of nowhere so differently?

"Do you worry that minors just don't understand enough about themselves—they're not neurologically developed enough yet to make permanent, life-altering decisions?" I asked Dr. Bowers.

"Absolutely not. No. Absolutely not," he quickly retorted. "As a sixteen/seventeen-year-old, have I changed attitudes and made mistakes from back then and done crazy things that I never would have done if I hadn't had the benefit of time? Absolutely. But when it comes to a core principle like gender identity, oh no. Gender identity is a core essential. That is, like, who you are."

Is one's identity in and of itself a "core essential?" Or is that status only conferred to gender identity? I probed further.

"This might be a bridge too far, but I don't know if you ever heard of people in the trans-abled community," I ventured. "These are people who are physically able-bodied but feel like they should be disabled, or identify as such. For example, a man who has two arms but feels like he should have one. Do you think that if a man in this kind of marginalized community went to a doctor and said 'I want to have my arm cut off,' do you think that..."

Dr. Bowers interrupted me. "That doesn't have anything to do with gender identity. I mean, that's a conservative argument, so I don't know where you're coming from." Obviously in the trans world, arguments deemed "conservative" are prima facie illegitimate. He continued: "That sounds like the apotemnophilia group that tries to liken, you know, having a penis off with, you know..."

Apotemnophilia is a condition where a person has an overwhelming desire to amputate one of their limbs.

"That's skunky. You know, that's a skunky argument," Dr. Bowers said.

I was confused. I couldn't see the distinction. So I asked, "What's wrong with it?"

"It doesn't have anything to do with gender identity," he shot back.

Obviously he wasn't seeing the root of the question, which is why gender identity is treated differently from every other type of proclaimed identity.

I tried to clarify, but he continued. "I'll accept it [apotemnophilia] as a mental diagnosis. A psychiatric condition... Somehow it's the idea that, you know, you're fascinated or charmed by having a limb or part of a limb missing. Ok, I would say that is, uh, pardon my nonmedical language, kooky. And I suppose that's a subjective assignment, but I'd call that kooky."

Kooky. Huh. Some people might describe sex changes that way.

Not to mention, maybe Dr. Bowers forgot—or wasn't even aware—that gender identity disorder was not that recently considered a "mental diagnosis" and a "psychiatric condition." Feelings of gender dysphoria were only elevated from psychosis to identity in 2012 and 2013 when the *Diagnostic and Statistical Manual of Mental Disorders* was updated in an act of political will by a small cadre of ideological psychiatrists.

"You think this is totally irrelevant?" I asked.

"Yep," he responded quickly.

"This is a man who says, I don't think I should have a penis. It's an appendage of the body."

I didn't want to let up. Dr. Bowers didn't either. He answered with a story about a man in his twenties who came in, who identified as a man, wanted to stay socially a man, wasn't feminine, and wasn't on hormones but wanted sex change surgery because he "thought [his] penis was just so ugly that [he] really couldn't live with it." Dr. Bowers counseled him for years and tried to convince him not to go through with it, but "eventually I did the surgery and I created a vagina," Dr. Bowers told me. "He could not be happier."

Dr. Bowers added—unsolicited, I might add—that he agreed the man's former penis was found aesthetically wanting. Actually, he put it much more bluntly: "That was the ugliest freaking penis I have ever seen."

I'm not sure why we had to go there, but oddly enough his story proved my point. If that young man can get surgery and mutilate his genitals even when it has no relationship whatsoever to his gender identity—and Dr. Bowers affirmed his decision in the most direct way possible by actually performing the surgery—then isn't that exactly the same as someone who wants to cut off his arm?

"I don't think so. No, I don't think so," Dr. Bowers said. The difference all seemed to come down to aesthetics. "That's like saying,

I got this huge, huge mole on my arm, and it's so ugly and there is hair growing out of it, and I'm, like, you know, I think any cosmetic surgeon or dermatologist would say that mole's gotta go."

But what if a person's arm is horrendously ugly? Well, it's not the ugliness that matters; it's the waste, according to Dr. Bowers. "I didn't cut [the penis] off. We repurposed it. Everything is used… all the components are still used… We're not going to chop an arm off that's healthy and throw it in a can."

Well, what if the amputee repurposed his arm?

"Are you going to make a planter out of it?" Dr. Bowers said. "Find another surgeon."

It was just like my conversation with Comfrey about my son's identity as broccoli. Both Comfrey and Dr. Bowers essentially just declared that gender is different than any other kind of identity. It is an assertion, not an explanation. There's no logical basis for the assertion, but it's an assertion that, if repeated enough, works. If you say it enough, people actually *will* believe that gender is different from every other kind of identity.

"[Young people] are being taught that [gender is just what you think you are] as if it's dogma; it's fact—two times two is four, the capital of the United States is Washington, D.C., you might feel like you're a boy, even if you have a vagina and you are a girl, you are what you feel you are," Dr. Grossman explained. "So they're being taught this as a package. They're taught that this is reality and that this is what the authorities are teaching—all the doctors and all the nurses and all the teachers."

The moment you question that dogma, you are no longer affirming. You are shunned. You can be labeled a hater and a bigot. You can lose your daughter forever. As Abigail Shrier wrote of parents whose young girls experienced rapid onset gender dysphoria, "they are terrified their daughters will discover the depth of their dissent and cut them off." The kids have their friends. They have an online cacophony of support and affirmation. They have affirming therapists and doctors and counselors. Why would you need a parent who tells you "no"—a parent who denies *who you are*?

The result of this mass social pressure campaign is measurable. In 2014 there were twenty-four gender clinics in America, almost all in California or along the east coast. The next year, there were forty. Not long ago, the idea of "affirming care" didn't exist. Now, there are 215 gender-affirming pediatric residency programs pumping out future pediatricians who will take every proclamation of gender dysphoria at face value and start helping children to transition.[22]

Transgender ideology was created and propagated by previous generations, from doctors in Money's offices in Johns Hopkins to writers and producers of the silver screens of Hollywood. Now, it has been taken up by the current generation, which is leading the charge in full force. Where are they leading us?

"That's the cool thing about gender," Dr. Forcier told me. "The quest continues for all of us, and the quest continues for the next generation, who's already telling us that our antiquated ideas of things have to be a certain way just don't apply to them. And that's exciting."

Dr. Bowers spoke a similar way. "You know who gets it right away is the next generation. You know, they're starting to figure it out. They're talking about it. They're exploring it. They're rejecting a lot of our social mores and these somewhat arbitrary rules."

But it's about so much more than the ideas and mores that they are tearing down. The next generation is charting a path to a new world that transcends the old. "There has long been a minority of people who would like to roll back the clock and have everyone have [Barbara] Billingsley as their mother making peanut butter cookies every day when we came home from school," Dr. Bowers contended. "But that's not progress. And then there are others who actually are looking for a better world, who are looking to move towards a place where there can actually be universal happiness, and of a planet that is looking out after every individual and every creature, and where human happiness is really a possibility."

"Progress." "Universal happiness." Paradise on earth. It's the goal of revolutionaries in every age. The ideals and battle cries were different in the past—liberty, equality, fraternity; the triumph of the proletariat; the brotherhood of man—but the desire was the same, the desire for mankind to finally be at peace with each other and at peace within himself. These new revolutionaries believe they've found the answer that escaped those in the past: True happiness comes from acceptance, from affirmation, from the self. It comes from my will and my understanding and my identity.

But the self is not enough. The self by itself cannot satisfy—especially when the world refuses to affirm "my truth." So reality must be altered to fit the self. Words must change. Clothing must change. Culture must change—and it all has. But the body—nature—resists. So the body must be made to submit as well.

Happiness demands sacrifice. Progress demands the shedding of blood. This revolution, like all others, comes at the point of a knife—perhaps not on battlefields, but certainly in surgical centers.

THE PROMISE OF TRANSITION

"ANY COUNTRY REPORTING data has shown that hormones and surgery are effective. They do work. Psychological well-being is better. Social standing is better. Grades improved. People get off their antidepressants. They quit or they lessen their use of anti-social and self-destructive behavior, and they get off drugs and alcohol more likely. They're happier people. They're more well-adjusted. It's overwhelming. That's case closed. That's done. It's not even a point of discussion anymore."

Dr. Bowers was definitive. The benefits of medically transitioning are undeniable. Full stop. Period. Science has spoken.

He wasn't the only one to wonder why I even asked the question. The media reports are just as definitive. Puberty blockers are "fully reversible." Taking testosterone or estrogen is affirming. Sex change surgeries reduce suicide and improve mental health.

Therapist Gert Comfrey and Dr. Michelle Forcier were likewise categorically confident. I may as well have asked them about their thoughts on other obvious questions, like if they believed the sun was hot, if they knew that the earth was round, or if they thought Trump was evil. The goodness of transition was foundational, and you can't waste your time talking about such first principle.

But I was here to learn. So ask I did.

THE ONLY MEDICAL TREATMENT WITHOUT SIDE EFFECTS!

"I know for other transgender people, there's like a spiritual component, like being on hormones and having hormone levels align deeply with who people know themselves to be is an incredibly spiritual experience," Comfrey told me. "I mean, I have a friend who every time they do their hormone shot, like, every week is just, like, a ritual. It is a profoundly spiritual experience to be able to practice this type of bodily autonomy. Yeah. To be in the body that they want. And that feels right to them."

Some people go to church. Some people commune with nature. Others get an estrogen shot. Spiritual transcendence: this was even beyond what the medical literature preached.

Dr. Bowers added more. "I can personally tell you that having done 2,200 of these surgeries, I can think of three—and I can't even think of one in the last ten years—who genuinely regretted going through surgery. And two of those three people did that because they couldn't find romance in their new gender. That was all it was. The other one couldn't find a job."

Two thousand and two hundred cases and not a single person regretted the surgery in and of itself. Effectively a 100 percent success rate. I've never heard of a medical intervention that was so overwhelmingly popular.

Dr. Forcier emphasized that medically transitioning doesn't just make people happy with absolutely no regrets. It is also safer than almost anything else you can do in life. "When a kid says puberty is hurting me, and I have a medication that is reversible," she told me, "that has no permanent effects on their gender and fertility and so many things that parents worry about, I have a medication that we can apply which is safer than Tylenol."

"How do you know that?" I asked.

"Because people don't OD on puberty blockers, and people OD on Tylenol, and they ruin their livers."

Who knew? I wonder why puberty blockers aren't available over the counter then and why Tylenol is so common. I marked that as something to look into further.

To be fair, both Dr. Bowers and Comfrey acknowledged that medical transition is not totally and completely sunshine and rainbows.

"Surgery has consequences. Hormones have consequences," Dr. Bowers said. And that's all he said. What those consequences are was left unstated, so I presumed they must not be that bad.

Comfrey struck a similar tone. "I think there's always risks, right," Comfrey acknowledged. "I think, like, again, not a medical doctor, but I know that, like, regular blood work is important for people who are on hormones to check things like, you know, how's your liver doing. Things like that." Hmm. I guess puberty blockers are easy on the liver and cross-sex hormones are hard on the liver?

Yet even Comfrey's admission of seemingly minor maintenance like blood work and checkups did have to be properly balanced with a description of what happens if transgender people forgo medical intervention. "It can have negative impacts if folks really need or want to show up in a certain way and are... denied access to that," she said. She cautioned me to "look at any type of surgery through the lens of, like, is this in the best interest of this whole person, yeah, not just reducing them down to, like, their body, but, like, yeah, their whole concept of self."

It was pretty obvious that, according to Comfrey, surgery and hormones may have some extremely minor physical side effects, but the benefit for the person's "whole concept of self" far outweighs any minimal costs.

I have to admit I was skeptical. Every medical intervention has side effects. Every time you hear a drug advertised on TV, right after they spend twenty-five seconds talking about how amazing it is, they spend five seconds going rapid fire through all the ways it can kill you. Was gender transitioning—puberty blockers and hormones and sex change surgery—really the only major medical intervention in existence whose side effects were so minimal that they barely needed to be mentioned?

All this information was coming from the same people who didn't seem to know where gender theory came from, struggled to differentiate sex and gender, and, in Comfrey and Dr. Forcier's case, definitively declared that there is no such thing as objective truth. How could they be certain on this? Why was everything else in gender fluid and a journey, but the science on medical transitioning was definitive and final?

I wanted to be sure. I wanted to follow the science. Transgender ideology may be logically inconsistent, as I found. The base ideas may come from really terrible and depraved people, as I learned. It may make a lot of people uncomfortable just how quickly transgenderism became the center of the national conversation in every sphere of life. But if it works, it works, right? These people I was talking with may not know what a woman is, but if calling a biological man a woman and making him look like he was "assigned female at birth" makes him happy, then who would I be to judge?

I decided to swallow my doubts one last time. After all, for Dr. Forcier and Dr. Bowers, medicine is their field. Dr. Bowers informed me that he is approaching the Guinness Book of world records for the most vaginoplasties, and that he conducts up to 200-300 surgeries a year. Dr. Forcier has been a doctor for LGBTQ kids in North Carolina for twenty years and works with gender hormones. And Gert Comfrey is a certified therapist—a master of mental health. All my other questions so far could be considered outside of their lanes of expertise. Now I was firmly in their lane. I decided to trust their proclamations one last time and try to find out with an open mind if medically transitioning really was a sort of modern-day miracle—a low-risk to no-cost way of making people truly happy.

But to get a better look, I had to get a remedial education in how medical transitioning and sex change surgery actually work. I wanted to start out by getting the dry facts.

THE PROCESS OF CHANGING SEX

Right off the bat, I was corrected. "I don't use the term 'sex change' anymore," Dr. Bowers told me. "I think that language is also a fluid, societal entity, and it should change and does change. Webster's dictionary changes with the times." Dr. Bowers said that "genital correlation surgery" or "gender affirmation surgery" are better terms.

Oddly enough, Dr. Bowers had explicitly used the phrase "sex change" four times in our interviews up to that point without making this linguistic correction. Generally, I find when someone all a sudden aggressively asserts the use of euphemisms, he's hiding something. But I decided to let it slide.

Dr. Bowers told me that he is "at the end of the food chain" after people have gone down a "long pathway" that is "evidence-based" and developed by "scientific consensus." "[Transitioning] isn't something that just pops up in your head one day and, oh gosh, wouldn't that be a great idea," he said. So who is at the beginning of that food chain? Generally it starts with therapy. It starts with someone like Gert Comfrey.

When someone believes they are born in the wrong body, they often find their way first in a gender-affirming therapist's office. The entire basis of affirming care is to not question the person's psychological disconnect from his or her biological sex, but to support and cultivate that disconnect. If gender is a journey, this is the start of the journey where a full medical transition—hormones and surgery included—is frequently the end. As a result, if there is dissatisfaction as a result of that disconnect, it is not the mind that must change but the body.

"Changing our bodies, being authentically in our bodies, is, like, mental health," Comfrey told me. "It's important in a mental health level to be able to practice that type of autonomy in our own bodies."

Once affirmed in this way, therapists like Comfrey can still remain in the equation as a mental health support, helping people process and get affirmation for every step they are taking. But the journey itself continues onward deeper into the medical establishment. "I'm not a medical prescriber at all, so I can't prescribe meds or anything like that," Comfrey told me. The type of person who can prescribe hormones, on the other hand, is Dr. Forcier.

Hormone therapy is a common intermediary step between initial psychological counseling and sex change surgery itself. "Trans people are seeking out hormones because they will have gender affirming effects," Comfrey told me. Biological women can begin to feel like men and vice versa through the impacts of these drugs. "That includes someone saying I want to be on testosterone because I want to experience having facial hair or I want to, you know, be on hormone replacement therapy because I want the effects of, like, breast growth," Comfrey added. "So a lot of it is around, like, these secondary sex characteristics that transgender people are seeking that feel really gender affirming. So it's, like, oh yeah, like, if I had this experience in my body, this will affirm and more closely align with the gender that I know myself to be."

Hormone therapy can come in several forms. For children, puberty blockers are used in an attempt to stop puberty and the normal development of the male or female body. These puberty blockers obstruct hormones and stop normal sexual and physical development. Perhaps the most common puberty blocker is a drug called Lupron.

The idea is that when gender-confused kids are afraid of developing a certain way, this alleged pause gives them time to choose if they would prefer to have medical intervention through cross-sex hormones or even surgery to force their body to develop in a chosen way—all before their body kicks into gear and develops according to the natural plan. Alternatively, Dr. Forcier told me puberty blockers allow children to stop puberty indefinitely and stay prepubescent forever.

Sometimes kids come to her, she told me, saying, "I need puberty to stop right now. I either need it to not continue, period, and I know that, or I need time to think about things and make decisions about what puberty might look [like] for me." Apparently puberty is optional now.

Those decisions about what puberty might look like are often made with the use of cross-sex hormones, which can change someone's physical development to appear like the opposite sex and can be administered via injection, cream, gel, spray, or a patch. Through this intervention, testosterone is used to make a female adopt the characteristics of a male, and estrogen to make a male adopt the characteristics of a female.[1]

It can take months for the hormones to begin to have an impact. For women taking testosterone, effects may include increased hair growth, male pattern baldness, increased muscle mass, an end to their period, and an atrophying of the vagina. For men who receive estrogen, the size of their penis and testicles reduces, they have a harder time getting an erection, they lose muscle mass, and they begin to form fat that appears like breasts.[2]

The use of cross-sex hormones in both men and women increases body fat and causes weight gain.[3] Additionally, cross-sex hormones must be taken for a person's entire life to maintain the appearance of the opposite sex, and that person must be regularly monitored by a doctor.[4]

Finally, many transgender people will decide to get genital surgery to construct the approximation of a penis, vagina, or breasts to fit their gender identity. That's where Dr. Forcier's world ends and Dr. Bowers's begins.

Dr. Bowers assured me that sex change surgery—ahem, gender affirmation surgery—is really very simple. "There is another term called sexual dimorphism, which is a biological term... and it refers to the amount, the degree, of chromosomal correlation with gender," he informed me. "In many species the amount of sexual dimorphism is actually quite large. In humans, 99.7 percent of the DNA is the same. So our sexual dimorphism between males and females is actually not that great in terms of our actual bodies and our chromosomes."

What does that all mean? Dr. Bowers clarified: "We're really not that different, males and females."

As a result, surgical manipulation of the body to change someone's sexual characteristics, in Dr. Bowers's mind, is not all that complicated. And that can be done in a rather wide variety of ways.

It's always easier to remove than to build, so we'll start with surgery for when men want to become women. This can include a wide array of cosmetic changes, from facial reconstruction plastic surgery to breast augmentation to something called a tracheal shave. This is where doctors shave down a man's Adam's apple to help him achieve a more feminine neckline. The most intensive intervention, however, is called a vaginoplasty—something that Dr. Bowers specializes in. In a vaginoplasty, the testicles are often castrated in a surgery called an orchiectomy. Then the penis is cut open, turned inside out, pushed inside a hole made in the body, and used to form something that looks like a vagina.

The entire process is physically grueling even after the up to six-hour-long surgery[5] is done. Patients are required to stay in the hospital for five or six days after the operation since they cannot walk or pass urine for days.[6] After leaving the hospital, recovery takes another six to eight weeks wherein the patient has limited mobility, largely can't drive, and is unable to lift more than five to ten pounds.[7] Despite Dr. Bowers's assurances on sexual dimorphism, the male body doesn't naturally accept the new opening made in its body, so regular "vaginal" rinses are used to keep the opening clean and machines are used to keep the gap from, well, closing. "So they have the dilators that are used to, uh, to, uh, to keep the bathroom walls open," Dr. Bowers told me.

I asked Dr. Bowers how long patients are required to do this... maintenance.

"We say indefinitely, really, which probably is true, although there is probably less maintenance after a year or two," he said.

As with hormone therapy, it appears that once you decide to undergo sex change surgery, you must spend your entire life convincing your body to do what you tell it. Nature won't ever come and assist.

So-called female-to-male transitioning is even more difficult. For would-be men, there is also a wide array of cosmetic surgeries to choose from, such as facial masculinization surgery, hair transplants, and breast reduction surgery. When it comes to what some call "bottom surgery," these women have more options than men who want to transition. They can get something called a scrotoplasty to create something that looks like a scrotum, and they also have two different options to mimic a penis. The first is called a metoidioplasty, where a woman is pumped with enough hormones to enlarge the clitoris and then that can be cut up and reconstructed to look like a penis.

The second option is called a phalloplasty, which reforms skin from the arm, thigh, back, or abdomen into the approximation of a penis. That new skin is then attached at the pelvis like a sort of biological strap on. More medical intervention is needed to make the approximation act something like the real thing. Because what is constructed is obviously not a real penis, women who have this surgery can get a penile implant that inflates to make the penis look erect.

The pro-transgender side may claim that puberty blockers are completely reversible, but Dr. Bowers was clear-cut on the effects of surgical transitioning, calling it a "permanent, irreversible

Rational from _Kasheena_

__ 1. Condition of the item

__ 2. Age of the item

__ 3. Duplicate copies

__ 3b. Newer editions on shelf

__ 3c. Newer edition available?

✓ 4. Outdated/inaccurate info

__ 5. Items with low/no usage

__ 6. Other material fulfills the need more adequately

__ 7. No longer supports the curriculum.

change to [the] body." When you remove your sex organs and try to insert others, there is no going back.

I'm sorry that I had to go into such detail. Honestly, I held a lot back. The play-by-play of each surgical process would make for rather heavy and gruesome reading. But even if I spared many of the gory specifics, I still thought it best that you knew what people are doing to themselves and to their children. When people say that transitioning makes transgender people happy, it's about so much more than pronouns and cross-dressing. Often surgery is on the table too. And the surgery is intense.

Now that I knew what transitioning entailed, I had a hard time believing that all of this pain and cutting and recon-structing was one of the keys to "universal happiness," as Dr. Bowers called it. Something wasn't passing the smell test. How could a biological man believe that being a woman is who he *really* is when his attempts to adopt the biological realities of a woman so clearly resist nature? His body treats the new vagina made in him like a wound that continually attempts to close unless he intervenes to stop it.

And how can a biological woman believe that being a man is who she *really* is when the penis she had constructed must be medically inflated to stay erect? Her body can't recognize the new attachment, so it must be made to do what most men have just about no trouble doing naturally.

Could these people really be so happy in such open rebellion against their bodies? Are our bodies really just fleshy bags of water to be manipulated as we will? On the first question, all the doctors and experts were telling me the answer was an un-equivocal "yes." The pro-trans experts I had spoken with never even thought to ask the second question.

WHAT ARE WE ALLOWED TO AFFIRM?

It seemed that either my gut reaction was completely wrong, or somewhere along the way the pro-trans argument broke down. I retraced my steps back to the beginning. While it's not a hard and fast process, transitioning follows a fairly regular pathway: first is affirmative therapy, then it's hormone treatments, then it's surgical transitioning.

I decided to question the very first assertion—the assertion that every pro-trans person made so definitively it seemed almost insulting to insinuate that they might be wrong. That is the idea that affirmative care is clearly the only and best way to interact with transgender patients.

Dr. Forcier made the point crystal clear: "When we don't listen, when we don't tell kids you're perfect just the way you are, they're not going to do as well as the kids that get 100 percent affirmation." One hundred percent. There is no questioning anything about a child at all.

Then it dawned on me; I have never met a single, successful parent that did anything like this. What kind of parent tells their kids they are perfect just the way they are? Why send your kid to school? Why tell him to sit up straight? Why show them how to share their toys? Why teach her not to lie or cheat? Parenthood is one long, drawn out process of continually molding and shaping your kids to be something better because they *aren't* perfect just the way they are.

That doesn't mean we don't love our kids. But love isn't the same thing as affirmation. Love isn't the endless repetition of "yes." Love is frequently a series of "no." No, you can't keep eating that. No, you can't touch that. No, you shouldn't hang out with those people. Without those "nos" your kid will become an obese burn victim hanging out with drug addicts.

Dr. Miriam Grossman addressed the difficulty parents especially have resisting the draw of completely and totally affirming their children. "'Affirming treatment... sounds so positive and so wonderful.... Why would anyone not affirm? It's wonderful to affirm somebody. Well, but what you're affirming here is a young person's self-perception, which, by the way, may not be clear at all. They may be very confused about it. One day they may be clear, the next day they may be totally confused and not sure."

I brought this idea up with Dr. Forcier—the idea that affirmation can't just mean saying "yes" to anything. Yet again she asserted inexplicably that gender identity is in a different category than everything else in human existence.

"We're going through a phase after we went to the zoo where my son says he is a spider monkey," I said.

"Are we going to do that again?" Dr. Forcier responded. "Are we going to do the chicken [example]... as a way to disparage people from thinking about the world?"

"Why are you disparaging him right now? I'm telling you how he identifies," I answered back.

"A young person talking about being a spider monkey is different than a person exploring their [sic] gender identity."

"You just said you have your own truth, right?"

"Mhm"

"That's his truth."

"Yeah, but that's not gender identity."

"I know it's a different thing, but we're drawing an analogy," I said, masking my frustration at the circularity of it all.

"No, the analogy doesn't fit," she said. "Because the analogy, again, is you're trying to sensationalize and pathologize what thousands of patients know to be their truth and their existence."

Yet again she refused to answer why gender identity must be respected and affirmed no matter what, while other identities can be scoffed at and disregarded.

"Let me ask you: How many transgender patients have you cared for or lived with or helped out at a street shelter?" she asked me. "How many transgender persons and families of transgender persons have you interacted with over the last twenty years?"

I admitted I haven't treated any transgender patients, and my interactions with transgender persons were thus far limited. But then again, how many six-year-olds has she treated who identify as spider monkeys? Appeals to interpersonal experience may matter when it comes to empathy. But they only go so far when it comes to facts.

That being said, maybe she had a point—not about the facts of identity or treatment, but about how well I can understand transgender issues without talking deeply with an actual transgender person. So that's what I decided to do.

THE TRUE STORY OF A TRANSITIONER

As I sat across from Scott Newgent in a beautiful room in New York with the warm sun showering in, I thought she looked almost like a middle-aged man: stubbled cheeks, thinning hair, a portly body shape. She didn't have the appearance of a woman, but neither did she have the full appearance of a man. One thing I noticed about all of the trans people I interviewed is that none of them could totally escape the biological confines of their true sex. They all ended up in a sort of gender purgatory.

Newgent lives in Fort Wayne, Indiana, but flew out to New York to meet with me while I was in town. I later learned that she hadn't left her house in three years. But there she was, eager to meet with me. That's how strongly she felt about telling her story.

I started off with the basics. "What led up to your decision to transition?" I asked.

She began with a sigh. "Well, it's a question I get a lot."

She told me about how she is a lesbian who was deeply in love with a conservative Catholic girl. The girl knew she wasn't a lesbian and told Newgent she acted like a man in a woman's body. "So I thought about it for a long time," Newgent said. "And it came to me that if I replaced my life, if I replaced my chromosomes from female to male, my life would have been completely different. I would have been the ultimate male. I mean, I would have been the football quarterback. All the movies that you see, that would have been who I was."

When she started thinking like this, she began seeing signs everywhere. "I never fit. I was an alpha female, a sales executive that kind of just didn't fit in any box." Her psychologist and people around her continued to seed the idea that she really was born in the wrong body.

"I started to think, 'Well, maybe I am.' It just kind of went from there," she said. "You know, my first therapist session, my therapist said, 'When did you start dressing like a male?' Now, I never dressed like a man. I was never a butch lesbian. I came from a very feminine family. I was in business sales and heels and all that stuff. Of course, I didn't have tons of makeup on—it was the end of the day—but I still had earrings on, and I looked very feminine."

But the insinuation pierced her. Because she was in such a vulnerable place and so open to suggestions, she thought her therapist must be right about her clothing choices and she must be wrong. Maybe she wasn't really feminine after all. "I kind of looked down, looked up, and went, 'I guess my whole life!'" she said, imitating her earlier surprise.

All of a sudden, her experiences as a lesbian and being uncomfortable started to fit together. At forty-two, she made the plunge. She decided to medically transition. I presumed now she considers herself a man. I wanted to ask her about that experience. But then she said something that surprised me. She spoke with a simple clarity and conviction that I never heard from Dr. Bowers, Dr. Forcier, or Gert Comfrey.

"I'm a biological woman that medically transitioned to appear like a male through synthetic hormones and surgery," she declared. "I will never be a man, ever."

But isn't gender a social construct? A journey? Gender identity is different from every other form of identity. It can change, and that must be respected and supported and affirmed. That's what everyone told me.

"Medical transition is an illusion; you create an illusion of the opposite sex—there is no such thing as changing genders. You can't," Newgent continued.

I was getting a little nervous. If I had said anything like that to the pro-trans crowd, they would turn on me in an instant. "Isn't this transphobia now against yourself, maybe?" I asked tentatively.

"No, it's reality. It's reality," she said. "The reality is, is that if you medically transition, you create an illusion of the opposite sex for comfort. Why is that transphobic?"

But then what is transgender surgery if it's not affirming someone's real, actual identity? "It's no different than if a woman has size B breasts, and she goes to a plastic surgeon and gets double D's and she's running around the streets with her boobs hanging out, going 'I was born with these. These are mine genetically. These are mine!'" Newgent told me. "People would think she's nuts, right?"

So was I correct all along? Was the idea that you can change genders like the idea that you can be a spider monkey or be affirmed as a transabled person and cut off your arm? Maybe just because some people believe something—even if it's about their gender—that doesn't necessarily make it real.

In reality, Newgent says, transgender people aren't affirming an identity. They are adopting a persona. "Why would people with dark hair want to have blond hair? Why would people with regular lips want to have red lipstick? I mean, it's cosmetic, right?" Newgent said. "We've been doing that since the beginning of time. Why is medical transitioning any different?"

We may have new technology and go to more extreme lengths to achieve our desired image, but the base impulse is the same. Ironically, that was a lot like what Comfrey told me: "I do know that trans people are seeking out hormones because they will have gender-affirming effects. So that includes someone saying I want to be on testosterone because I want to experience having facial hair or I want to, you know, be on hormone replacement therapy because I want the effects of, like, breast growth."

Comfrey confirmed that the same pursuit of a particular type of image is what drives the desire for sex change surgery as well. "Surgeries are options for some people, a route that some people go down. And yeah, it's in the spirit of, like, affirming that person, affirming their body and, like, how they want to just show up in the world and in their own body."

Newgent seemed definitive, just as confident as the pro-gender theory people I had spoken to for hours before. But she was saying the exact opposite. We shouldn't affirm people in their transition because it contradicts reality. Getting a sex change surgery doesn't actually change your sex. You can't be a man in a woman's body. Everything I was being told before appeared to be built on a lie. "Maybe the whole agenda is to get everybody to think that being truthful is transphobic," Newgent said. That would shut down debate, after all.

USING SUICIDE THREATS AS MORAL BLACKMAIL

I hit the brake pedals hard. This was all making a lot of sense intellectually. But what would be the real world impact? The experts told me that if we didn't affirm everyone in their gender identity, then they would commit suicide. It seemed definitive. After all, Dr. Forcier told me, "If their families love and accept [LGBQIA++ rainbow children], they have reduced risk for depression, anxiety, suicidality. They just do better. Kind of a 'duh' statement."

Ultimately, I didn't want to believe in a lie. But I also didn't want to have blood on my hands.

Newgent quickly overturned everything I had heard before. "We find with children that have gender dysphoria, if and when they enter into talk therapy, 82 percent kind of grow out of that and grow into loving being a woman."

Dr. Grossman firmly backed Newgent up. "The claim that if we don't affirm every single child—affirm their gender, their per-ceived identity—that they will go on to hurt themselves and that they will go on to even perhaps kill themselves is not valid."

Newgent actually believes that affirmative care causes even more problems by pressuring kids who would otherwise have no psychological issues to embrace an identity out of sync with their sex—a disconnect that everyone agrees comes with great mental anguish.

"You know if we told middle schoolers right now," Newgent said, "at a time they're not fitting in, that you can absolutely fit in. The only thing that you need to do is cut your right leg off, and you will fit in for the rest of your life. Do you know how many ambulances we would have to middle schools?"

Dr. Grossman and Newgent weren't the only ones who spoke clearly and forcefully against the idea of affirmative care. During my journey, I had the opportunity to sit across the table from Dr. Jordan Peterson, an internationally renowned clinical psychologist famous for boldly speaking the truth and describing the scientific nature of how human beings think, act, and believe.

As a psychologist well-versed in therapy, Dr. Peterson was swift to condemn the very idea of affirmative care. "There's no such thing as a gender-affirming therapist. That's a contradiction in terms," he told me. "If you're a therapist, it's not your business to affirm."

He continued on, "I don't affirm what you're saying, that's for sure," he said. "That's not therapy. That's a rubber stamp. But when we're talking about something as complicated as gender and sexual identity, that's like complicated right down to the core of being. It's like, you don't get a casual pat on the back from a therapist for your preexisting axiomatic conclusions."

What is the point of going to a therapist then if they aren't going to affirm me in whatever I say or do? Dr. Peterson explained further. "Why do you come and see me? Because I'm going to listen to you. I'm going to listen to you, walk through your concerns, your problems, your goals, your ambitions, your dreams, your confusion, your hatred, your resentment... Maybe you come to see me because a destructive element of you is wreaking havoc in your life. I'm on the side of the part of you that wants to aim up."

I told him about my experience with Comfrey. "The therapist I talked to here was ready to affirm me as a woman because I said that..." I began, before he interjected.

"That wasn't a therapist," he said. "That was an ideologue, a terrified ideologue." All this made a lot of sense. Affirmation seemed like the completely wrong way to approach therapy. But it didn't change the fact that the studies Dr. Forcier and Dr. Bowers alluded to seemed to have a clear correlation: no affirmation = high suicide risk. Then I remembered a quote attributed to Nobel Prize winning economist Ronald Coase: "If you torture data long enough, it will confess to anything."

The truth is, those with gender dysphoria have roughly the same rate of suicide as youths with other mental health conditions.[8] They aren't exactly unique. That being said, the pro-trans side is correct on something. Gender-confused people who aren't affirmed in their gender identity are at a high risk for suicide. But the inverse is also true. Gender-confused people who *are* affirmed in their gender identity are also at a high risk for suicide. That's because people with gender dysphoria trend towards suicide no matter what.

Dr. Grossman unpacked it further. "These [gender dysphoric] kids, as I mentioned, have a lot of comorbidities. A lot of them were hurting themselves. A lot of them were suicidal before they even discovered gender. That is never part of the discussion when we talk about these kids hurting themselves and committing suicide."

People with gender dysphoria suffer from all sorts of emotional disorders, but in Dr. Grossman's experience, we interpret every experience they have through the lens of gender. There could be a million and one reasons why someone is unhappy. But the moment the person exhibits any sign or makes any statement or is open to any suggestion that he or she isn't comfortable with his or her sex, gender-affirmative care becomes the automatic answer. Every other possible cause gets pushed aside.

"What we're doing here is we're just putting it all in that one basket [of gender]," Dr. Grossman said. "This thinking is not clear. It's not. It's not backed up by the evidence that we have, and it's a way to emotionally blackmail the parents. They say, 'What would you rather have? A living daughter or a dead son?'"

That's true. That's not the first time I'd heard that line—and it all stemmed from the same flawed analysis of the transgender suicide data. Dr. Grossman continued, "If this is what the professionals are saying, I mean, my goodness, what parent is going to have the strength? Actually, there are parents now that are mustering up the strength to resist that. But it's terrible emotional blackmail."

It occurred to me that maybe the pro-trans side just wasn't keeping up over the long term with gender-dysphoric people who transitioned. Maybe they asked them if they were suicidal before their transition and compared it to right after their transition. Then, in the glow of the new, transgender people seem happy. They feel like they are finally at peace. But are they? I remember Dr. Bowers said that not a single patient has ever regretted transgender surgery in and of itself. But knowing what I know now, I found that increasingly hard to believe. What about Dr. Forcier?

"How many of [your patients] have you checked back on twenty years down the line to see how they're doing?" I asked Dr. Forcier.

"I have actually a number of patients over the last ten years I'm still in touch with from the program that I served," she said.

"I'm talking about all these kids who are getting the drugs now," I responded.

"Right, I can't—if they're getting the drugs now, twenty years is twenty years from now."

"So we don't really know [how they're doing]?"

"No, we don't."

Honestly, all of this was like determining how happy married people are by only asking newlyweds. Their answers could be different twenty years down the line. Heck, it could change after two! Except in this circumstance instead of getting married to someone you love, you get injected with hormones and have your genitals surgically mutilated in an irreversible way.

The stories of people who suffer from gender identity disorder don't start when they perceive a disconnect between their gender and their sex. And they don't end the moment they get affirmative care and start the medical transition process. These stories have endings—and the endings don't always fit the pro-trans narrative that I was being told.

It made me think of one troubling story I never saw through to the end. What ever did happen to the Reimer twins from the John Money experiments? I had learned a lot about gender theory since I was introduced to their terrible story. I know that David Reimer was never happy and transitioned back to being a male after all the lies the medical establishment had told him—lies I was starting to see are still being told today—that gender is a social construct and that gender identity is just some malleable journey.

But what happened to David Reimer and his family in the end?

"One quick follow up on the John Money story," I said to Dr. Grossman. "David Reimer. How did things work out for him? Is he doing well now?"

"No," she said heavily. "David Reimer ended up marrying a woman, adopting her three kids, and working as a janitor in a slaughterhouse. His story was told in this very, very important book, *As Nature Made Him.* Unfortunately, the trauma that he and his brother and his entire family went through left deep scars. His brother died of an overdose when he was thirty-eight. Then David committed suicide a few years later."

His parents blame John Money, the father of gender theory. They blame Money for the surgery. They blame Money for the insidious ideas he planted in their heads. They blame Money for his abuse of David and his brother year after agonizing year. Gender theory didn't kill David Reimer. That would be saying too much. But gender theory provided the excuse for inflicting so much pain upon him that David ended up taking his own life. What is gender theory doing to our children now? I would soon learn even more.

FALL OF THE HOUSE OF CARDS

DESPITE THE INSISTENCE of pro-trans activists that affirmative care was not only healthy, but also undeniably beneficial, it was becoming pretty clear that affirming gender confusion leads to serious complications and trauma. I had learned that affirmation leads to depression, anxiety, and even suicide, but what happens when people act on that affirmation and interfere with the body on a chemical level? Does it get even worse when they take the next, radical step of making surgical changes that can never be undone? I was about to find out.

"COMPLETELY REVERSIBLE?" YEAH, NOT SO MUCH

At first, I was trying to search for a silver lining somewhere. Thousands—if not tens of thousands—of people, including children, are gender-confused, and they are all being told repeatedly by culture and by therapists and by doctors that their sufferings will be relieved if they reject their biological sex and attempt to become someone else. They are being told the only way they can be happy is to set aside who they are and how they were born to become who they think they are.

Maybe it's not as bad as I thought. Sure, affirmative care is built on a lie, and it pushes people to embrace their confusion. Sure, it doesn't end with cross-dressing and new pronouns. But it could be worse, right? I mean, didn't Dr. Forcier tell me definitively that puberty blockers—the most common next step for young people after they begin socially transitioning—are completely reversible? I looked back in my notes.

"Puberty blockers—which are completely reversible and don't have permanent effects—are wonderful because we can put that pause on puberty," she said. "It's like if you were listening to music, you put the pause on. And we stop the blockers, and puberty would go right back to where it was. The next note in the song is just delayed that period of time."

"How do they do that? What actually do they do?" I asked her.

"This is how I would explain it to kids in a family. It's like a pretend hormone and they're going to go to your brain, and they're going to block the receptors in your brain so your brain glands that are sending messages to your gonads or ovaries or testes are going to think, 'Oh, my receptors are full. I don't need to do anything.' So the messages from our brain stop."

She continued, "Just like any medicine or other medication, they'll wear off if we don't give them another dose or replace the implant or stop it altogether. And you go into your puberty of your gonads and your assigned gender at birth, or you affirm with gender hormones that are more congruent with your gender identity. And those are all the options out there."

It seemed pretty straightforward. We shouldn't be pausing anyone's puberty—that's just another way we try to rebel against nature and affirm a disconnect from reality. But at least it's just a pause, right?

"So [hormone blockers] are totally safe. I was told, ok, they're totally safe..." I began relaying to Dr. Grossman what I had learned from Dr. Forcier.

"It's catastrophic that this is what you're being told," she shot back. "I just spoke a month or two ago with a mother whose fourteen-year-old daughter was put on blockers two years earlier. She had the hormone inserted under her skin on her arms so she wouldn't have to get shots. And they discovered after two years that her bone density had gone down by 30 percent. She has osteoporosis. This fourteen-year-old girl has osteoporosis! That's something that old women get. That means that her bones are brittle and that she could fall and easily break a bone."

Wait... but I thought puberty blockers were safer than Tylenol? Maybe this is just a short-term side effect.

"Is that reversible?" I asked Dr. Grossman.

"No, you can take... you can do stuff for it. But ask any sixty-five-year-old woman. 'Is your osteoporosis reversible?' And she'll say, 'Well, no, my doctor says it's not.' A fourteen-year-old girl!"

I went back to Scott Newgent, the transgender woman who had gone through an entire medical transition. If anyone could confirm whether Dr. Forcier was right or Dr. Grossman, she could, seeing as she's been through all of this.

"At a time when our timeline in our body tells us, 'Hey, it's time for your balls to grow, it's time for your penis to grow, it's time for your brain to grow' ... we're going to say it's OK to skip that and then come back to it?" she told me. "Our body doesn't work like that. So, we're seeing eighteen-, nineteen-, twenty-, twenty-one-year-olds with hearts and lungs the size of an eleven-year-olds."

What about cross-sex hormones? Those can't be as bad as puberty blockers, right?

"We do know that if males take estrogen for an extended period of time, it causes bone loss," Newgent informed me. "Try to find trans women that are over sixty-years-old, that have been on estrogen for over thirty years. Try to find them that are not in walkers, walking with canes."

But what about all the studies? What about the science?

"I've been told many times that they know this is perfectly safe," I told Newgent.

"Who does?" she asked quickly

"They."

"Who's they?! Where are the studies? Is that hidden right behind the unicorn farts and glitter bombs?"

The truth is, nobody on the pro-trans side ever really dug into the studies. They could cite a million headlines, but I didn't know if the data was legitimate. Maybe Newgent was right, and the entire structure of hormone therapy was actually built on nothing.

"There are no long-term case studies on hormone blockers, on synthetic hormones." Newgent went on, "How many studies do they have—long-term studies on hormone blockers for children? None. So we're saying 82 percent of children will recover from gender dysphoria. We're going to put them on hormone blockers. We don't know what the f*ck they're going to do to our children."

I wondered how doctors could prescribe drugs to children if it has never been studied. I asked Dr. Grossman what the hell was going on—and she confirmed everything. "[Doctors are] affirming it with hormones that have never been used in this way in the field of medicine."

What we do know, according to the *Journal of Clinical Medicine*, is that the long-term use of cross-sex hormones increases the risk of heart attacks, bone damage, liver and kidney failure, and pulmonary embolism. The American Heart Association adds that cross-sex hormones can cause blood clots and strokes as well. Newgent previously wrote about the effects of this therapy: "Almost a quarter of hormone-therapy patients on high-dose anabolic steroids (such as the testosterone taken by female-to-male transitioners) exhibit major mood-syndrome symptoms. Between three and twelve percent go on to develop symptoms of psychosis."

Yet even so, cross-sex hormones are so common that people can walk into Planned Parenthood and get the drugs that very same day. When used by adults, the effects can be terrible. But the real tragedy comes when hormone therapy is given to children.

As I learned before, perhaps the most common puberty blocker drug is called Lupron. It only took the most cursory search to confirm what Newgent and Dr. Grossman had said—a search I hadn't even thought to do since all the other doctors were so confident. Not only is Lupron not approved by the FDA to be used as a hormone blocker for gender-confused kids, it causes a wide array of adverse health effects that nobody had told me about before.

One woman had to have surgery to replace a deteriorating jaw joint at the age of only twenty-one. She had degenerative disc disease and a chronic pain condition called fibromyalgia. Others had brittle bones, faulty joints, mood swings, headaches, even cracked spines and thinning bones. One needed a total hip replacement when only twenty-six years old. According to another report, the most common side effects people experienced were depression and anxiety—maybe not because of Lupron itself, but certainly as a result of the pain and hardship they underwent after having taken it.

But it's not like Lupron is new. It's been around for decades. It seemed unconscionable that we would experiment on kids but not experiment first in a lab. Newgent told me it all boils down to keeping the transgender transitioning machine running. If people knew the truth, Lupron would go out of business.

"They don't want to get it FDA-approved because medical transition would stop immediately across the board."

CHEMICALLY CASTRATING KIDS

However, long before Lupron was ever used to block puberty in adolescent children, it had another purpose: chemical castration. Lupron reduces testosterone in men much like removing the testicles would, thereby reducing sexual urges. It has been used by some countries on sex offenders and other sexual deviants to prevent sexual crimes.

I asked Newgent about the issue directly: "Is Lupron chemical castration?"

"Yes," she said without hesitation. "We're giving it to pedophiles, aren't we?"

I had to confront Dr. Forcier on this. After all, she prescribes puberty blockers to children.

"[Puberty blockers] reduce the production of testosterone," I reminded her. "So that's chemical castration."

"No. The gonads are still there," she said.

"Well, it's chemical castration because it stops the production of testosterone," I answered back.

"No," she said. "I mean, I think what you're trying to do is use sort of like exotic and titillating words like 'drugs' and 'chemical castration' to do what the media likes to do, which is create drama around this."

"It's pretty accurate," I answered. After all, I thought to myself, if describing a drug accurately creates drama, it is the drug, not the description that is dramatic.

"We're talking about kids that we have a medication that's safe, we have a medication that's effective," she said, repeating the lies I had heard before.

I dug in deeper. "It stunts growth; it affects your bone density and that sort of thing, doesn't it?"

"No, no, no," she answered emphatically. "We use puberty blockers in a variety of ways to impact height... Bone mineral density catches back up."

Tell that to the girl with osteoporosis, I thought. Tell that to the woman who had to get her jaw joint replaced when she was twenty-one.

"Again, these [are] media things that people want to hear how exciting or dangerous or bad..." Dr. Forcier continued.

I interrupted. "I'm not excited about it at all. I just, you know, feel like using the correct terms is important. So for, say, chemically castrating kids..."

"That's not a correct term," she interjected. "That's not a correct term for puberty blocking."

"I can look it up on my phone. I'm pretty sure if I looked it up..."

"You can look it up on your phone."

"...it says 'medical definition: the administration of a drug to bring about a marked reduction in the body's production of androgens and especially testosterone.' That's chemical castration."

"Gender is different than diagnosing a medical problem," she responded. "And so if you want to apply medical, pathologic, and pathologizing concepts, you're doing harm to kids. And I don't do harm to kids in that way. So we use 'puberty blockers.'"

I'm causing harm to kids with my language? What about the fact that she's causing harm to kids with her drugs? Dr. Forcier didn't disagree with the definition. Like Dr. Bowers, she just wanted to use a euphemism. She wanted to sugarcoat reality to make it palatable for children.

I wasn't going to let her off the hook.

"One of the drugs used is Lupron," I reminded her, "which has actually been used to chemically castrate sex offenders."

"You know what? I'm not sure that we should continue with this interview because it seems like it was going in a particular direction."

"You don't want to talk about the drugs you give to kids?" I asked her.

She didn't. Like so many on the left she didn't want to engage with my ideas. She wanted to attack me as a person and make me afraid to ask questions she doesn't want to answer.

"When you used that terminology, you were being malignant and harmful," she said. "And I would appreciate that you use words and terminology that doesn't [sic] demonize the process of listening to diversity. And I would appreciate that rather than trying to create a titillating message, we talk about science, and we talk about science in a way that's respectful to children and families who experience this. And what you're doing now is not that."

I responded calmly, yet firmly, "There are some who would say that giving chemical castration drugs to kids is malignant and harmful."

"There are some that say that. Are they transgender?"

Yes, I thought, remembering Scott Newgent.

"Are they medical professionals?"

Yes, I thought again, remembering Dr. Grossman. But who cares!

"What does that matter?" I asked.

"Because it's about the context," she said. "It's about the context of caring for a child and seeing the suffering that kids can have and that we know adults have had and that we know have led to poorer, poorer health outcomes for a number of transgender patients that have not had access to care, that have not been in affirmative home situations, that have not had medical professionals support them. There is overwhelming literature that not supporting and not providing gender-affirmative care to persons who identify as gender-diverse or transgender promotes harm."

There we go. We were back to square one. She couldn't deny that puberty blockers are chemical castration, only that we couldn't call it that. And when she was finally pushed to explain how she could rationalize chemically castrating children, it came back to the same debunked assertions that gender-confused people *must* have affirming care or they will be unhappy—studies that don't even account for the fact that gender-confused people are almost always unhappy, whether they are affirmed or not.

Yet again, Dr. Grossman hit the nail on the head: "Affirming-treatment involves very dangerous and I would say experimental treatments. Affirming-treatment leads us to think, like Orwell said, that language can be used for a purpose to change the way that we think." That's exactly what Dr. Forcier was trying to get me to do. She was trying to change the way I thought about chemical castration by calling it something else. She was trying to change the way I thought about using a chemical castration drug on children by calling me "malignant" and "harmful."

HORMONES DO NOT IMPROVE HAPPINESS

It would at least provide a silver lining if these hormone therapies improved happiness in spite of all of the side effects. But of the two main studies that evaluated the effects of puberty blockers on mental health, one found absolutely no improvement and the other only found extremely marginal improvement. Other studies were fundamentally flawed or also failed to show any positive impact on psychological health.

Dr. Forcier rejected the idea that hormone therapy fails to help patients, citing specifically the Jack Turban study of 2020 that purportedly showed that transgender adolescents who had access to puberty blockers were 70 percent less likely to commit suicide. In Dr. Forcier's words, the study showed that "transgender persons are healthier, safer, happier when they have access to trans-affirmative, trans-sensitive, and trans-caring medical services."

The only problem is that study is a crock too. The study was done online by a pro-trans lobbying group, and only 11 percent of the 3,494 respondents had actually received puberty blocking drugs—an exceedingly small sample size. Even then, this pro-trans-funded survey still found that while 58 percent of those people who didn't receive puberty blockers planned to commit suicide, so did 55 percent of those who *did* receive puberty blockers—not that that difference matters much, considering the sample population for that question was only eighty-nine people.

But it gets worse. According to one analyst, "the numbers [of those] actually attempting suicide in the last year were higher in those who'd taken puberty blockers." Among those who were hospitalized after attempted suicide, nearly twice as many were *on* hormone blockers.

The study Dr. Forcier cited was paid for by a pro-trans group, statistically suspect, and not only did it fail to prove her point, in many ways it actually directly contradicted what she was telling me. Dr. Forcier was selling me a bill of goods, but I did have to thank her. I learned a valuable lesson. When the media and the pro-trans side cite a study proving their ideology, *don't believe any of it.* Dr. Forcier was either outright lying to me, or she never looked at the study herself. The headlines confirmed what she already believed, so she took it at face value.

Some are waking up to the lies. The United Kingdom's National Health Service (NHS) will now limit puberty blockers for children after a court found that children at such a young age can't possibly weigh the long-term consequences, and that research found that the impacts of puberty blockers on things like bone-mass density and height are not actually fully reversible. Quietly, the NHS removed language online claiming that puberty blockers could be reversed.

Finland, once an early adopter of medical transitioning for adolescents, also found that transitioning failed to improve mental health outcomes. In 2020, the Finish National Gender Identity Development Service recommended psychotherapy as the primary treatment for young, gender-confused people, not sex reassignment.

In Sweden, many hospitals and clinics have stopped medical transitions for those under eighteen because of the mounting evidence that medical transitioning has adverse side effects without improving psychological well-being.

Heck, there are even studies that reveal social transitioning—without any sort of medical intervention—fails to improve mental health because it pushes young people to persist in their gender-related distress when so many would have otherwise grown out of it.

Britain, Finland, and Sweden aren't exactly hotbeds of transgender oppression. They are simply following the facts. Yet in

America, hormone therapies and medical transitioning are sacrosanct. We aren't allowed to question it without medical professionals like Dr. Forcier turning on us with bitter anger. In fact, while other nations are pressing the brakes, our medical establishment is running full steam ahead.

When I asked Comfrey about how parental consent factors into these monumental medical decisions, her response was that "there's some wiggle room around, like, age of consent [in Tennessee] for, like, certain medical procedures. But yeah, for the most part it's, yeah." Dr. Forcier was even more expansive: "Medical affirmation begins when the patient says they're [sic] ready for it." If a patient is "ready" at nine years old, so be it, I guess.

Even discussing actual facts and data is not allowed if it risks contradicting the party line. In 2021, the Society for Evidence-Based Gender medicine wanted to share information about the impacts of gender-affirming treatment at the annual meeting of the American Academy of Pediatrics (AAP). The AAP rejected the application without any explanation—they refused to even consider data that could contradict the pro-affirmative care side—despite the fact that days earlier 80 percent of pediatricians at the meeting called for more caution when it comes to gender transitions for minors.

BOTCHED SEX-CHANGE HORROR STORY

But hormone therapy is far from where the horror story ends. Hormones are often only an intermediary step. Surgical transitioning is the coup de grâce.

Newgent described to me how the entire process of medical transitioning sucks people in a long downward spiral in search of the happiness they were promised.

"We're taking our most vulnerable kids in the entire world, and we're telling them that there's a fix for it... Because here's what happens with medical transitioning," she told me, drawing from her experience. "You start with the idea that 'I was born in the wrong body. Thank God, life is going to get better now. Right?' So we start hormones and then six months later after hormones, we go, 'Well that didn't help anything.'"

But these people have already gone so far. They can't turn back now. The only option they feel they have is to continue. Everyone in authority is telling them if they aren't happy yet, it's because they haven't been affirmed enough.

Newgent went on: "But I still need top surgery. So you get top surgery and then you go, 'OK, well, I still have that inner thing, you know.' I still need bottom surgery and I need to change my pronouns and I need to do this, and society's being transphobic because they're misgendering me, and there's always some kind of connection, right? Well, at some point, you kind of got to look left and right and go, 'Well, that didn't fix a thing!'"

Newgent's story arc is matched by the data, as nearly 100 percent of children who begin puberty blockers will proceed to cross-sex hormones and surgeries.

Dr. Bowers told me effectively no one regrets getting a sex change surgery. I didn't have faith that he was telling me the truth. Everything else the pro-trans ideologues told me had come toppling down like a house of cards. Surgical transitioning was the summit—the highest and most invasive form of affirmation. I had a hard time believing that sex change surgery worked when everything else had already failed.

Dr. Bowers had already called sex change surgery a "permanent, irreversible change." I asked Newgent what that change meant for her. The pain flooded out.

"I've had seven surgeries. I've had one stress heart attack. I've had a helicopter life ride with a pulmonary embolism. I've had seventeen rounds of antibiotics. I've had a month of IV antibiotics. I had a surgeon who was banned from conducting surgery in San Francisco, who moved to Texas, where they have a tort reform act where basically suing somebody with an experimental procedure is kind of slim to none, who used the wrong skin to create my urethra. I had six inches of hair on the inside of my urethra for seventeen months. I didn't sleep for seventeen months. I lost my job, my house, my car, my wife, everything I've ever worked for. And nobody knew what was wrong with me."

I didn't know where to start. She put her body through hell. How could this have happened?

"Medical transition is experimental," she told me. It's not regulated. It has been refined. But nonetheless a lot can go wrong. She tried to help me understand the depth of her agonizing, physical pain. "[The doctor] used the wrong side of the skin to create my urethra," which is the tube that carries urine from the bladder out of the penis in a male. "Think about having an ingrown hair on your face. Now, think of that with urine passing it on the inside of your body—moving and changing with puss, getting infected."

Every movement, every twitch, every time she had to go to the bathroom was drenched in pain. "I was so sick. I'm still sick," she added. But the physical pain was only part of the equation. She had spent so much of her money getting the surgery in the first place that she needed help fixing what the surgery had done.

"I got a job for three months—I don't know how—because I had to get insurance, because my mind wouldn't think. Get a sepsis infection and [imagine] how you could think. I moved across the country to figure out how to get somebody to help me."

I wondered why the doctor who caused the problem couldn't fix the problem. That's when she told me, "Nobody would help me, including the doctor that did this to me, because I lost my insurance. I worked for three months to get insurance until that insurance kicked in." Altogether, her medical expenses to both her and her insurance exceeded $900,000.

That wasn't the only problem. Because her issues were so particular, she couldn't go to any old doctor. Few are familiar with the intricacies of sex change surgery. "I had to go outside of the state that I was living in because nobody in the state knew what they were doing," she told me. "I had, and still have, and will always have a recurring infection for the rest of my life. At some point, antibiotics are not going to work anymore."

I looked deeply at Newgent, realizing I was talking with a woman who lived with a death sentence. "I get infections every three to four months," she went on. "I'm probably not going to live very long."

What about the impacts on her mental health? Obviously, the physical anguish and financial hardship were hard enough. But was there any improvement in her mental health by having her body "match" her gender identity?

"The thing is, is that most people that get phalloplasties have major PTSD. I haven't left my house in three f*ckin' years, and I'm here in New York. For three years."

Newgent's ordeal was tragic, but maybe just a tragic fluke.

"Is your story rare or common?" I asked. "Because I've been told that what you're talking about is an extreme outlier."

"It's not," she informed me. "Why don't you go to my website?"

I did, and I found testimony after testimony of people who regretted their transition—the exact type of people Dr. Bowers told me didn't exist.

"We have a gentleman... in Canada, Aaron Kimberly," Newgent continued. "He had the same surgery by the same person... Why don't you ask him how many times he saw it happen? Rare? It's not rare."

I brought up the alleged reams of evidence supporting medical transitioning and that Dr. Bowers told me without any hesitation... that "any country reporting data has shown that hormones and surgery are effective." She told me after transitioning people are happier and better off and well-adjusted—and that it's "not even a point of discussion anymore."

Newgent responded: "The only long-term study tells us seven to ten years is when transgender people are the most suicidal after surgery."

But what about all the other data? There has to be contradictory data right? Dr. Bowers and Dr. Forcier didn't just pull their propaganda out of a hat, I presumed.

Newgent admitted the studies are out there. But that's hardly the end of it. "We have studies that said that medical transition helps mental health, helps mental health of kids," Newgent said. "They've all been retracted, modified, changed. But you're not covering it. ABC's not covering it, NBC's not covering it because they're afraid."

How does she know they've been retracted if the media isn't covering it?

"In the UK, the attorney there that got [hormone blockers] banned [for minors] in the Keira Bell case—do you know how he did that? Because every time they threw down a study and said,

'Look, hormone blockers, they're safe.' Yeah, that one's been retracted. 'Look, it helps mental health.' Yeah, that one's been retracted." She went on, "and finally, the judge said, 'Do we have any studies that have not been retracted that it helps mental health?' No. OK, we're not going to talk about that. Move on."

As a result of this decision, in the UK, puberty blockers and cross-sex hormones were only to be administered to someone under sixteen if a court authorized it. The victory for children was short-lived, however, as another court soon overturned that decision.

Dr. Grossman confirmed everything Newgent just told me about how medical transitioning doesn't help mental health. "When we look at the statistics of adults that are years following their transition—their chemical, their medical transition, including surgeries—we see that they're still committing suicide at a remarkably higher rate than most of the population"

It takes only a little common sense to figure out why all these medical interventions just aren't working. Unlike other treatments and surgeries that address some real, physical problem to heal the body, transitioning harms the body to address a psychological problem. But people can never be at peace when they rebel against their nature. They certainly can't be happy when they inflict harm upon themselves during that rebellion.

A dress, a pronoun, a new name, a haircut, a hormone blocker there, an injection here, a cut there, a prosthetic here—the litany of illusions and lacerations from the social to the medical heaped one upon another can never change the immutable fact that every single cell in the human body screams "male" or "female."

This isn't conjecture. It's science.

Dr. Grossman described the process we all learned in school but somehow forgot. "How does that man get a man's body? He gets it as a result of his chromosomes, his Y chromosome," Dr. Grossman reviewed. "At conception, when the egg is fertilized by a sperm, the resulting organism either has two XX or an X and a Y [chromosome]. The presence of the Y chromosome is going to direct the development of that fetus in a male direction, not only in terms of his genitals, in terms of his brain as well... That means that in utero his body was masculinized, including his brain. What happened was that at eight weeks after fertilization, his Y chromosome sent out the instruction to his testes to create testosterone, and that testosterone was then distributed throughout the body."

This isn't some arbitrary process—and it can't be easily altered later in life. The impacts of this sex-specific development can be seen in other aspects of medicine far outside of "gender identity."

"Let's say a woman needs a kidney transplant and she gets a kidney from a male," Dr. Grossman posited. "So each of those cells in that transplanted kidney has a Y chromosome, and her female body can recognize that as foreign. Her female body doesn't recognize the Y chromosome. It never had a Y chromosome. So it's best to give a female kidney to a female." Ironically, at the same time gender theory has spread like a cancer, the field of sex-specific surgery has also exploded.

"So we have these two things happening at the same time in history," Dr. Grossman continued, speaking of research into sex-specific care and the rise of gender theory. "While the libraries, the medical libraries, and the journals are just filling up with articles indicating the vast and profound differences between male and female and how we must recognize that, we have, on the other hand, this ideology that is pushing false notions at us and at our children, and I don't know how much longer this can co-exist."

Dr. Jordan Peterson attributed this terrifying and unscientific medical experimentation to the very foundational claims of gender theory. It's not that gender theorists were wrong to want to differentiate between people's biological sex and their outward or psychological characteristics. It's that they invented the idea of "gender" with all its implications when we already had a perfectly fine way to understand what was going on.

"The issue with sex and gender, a lot of that's just deep ignorance and confusion, with trifling malevolence thrown in there and some willful blindness," he told me. "Biological sex [is] binary. It's been binary for, like, one hundred million years—longer than that, since sex evolved a long time ago. There's not three sexes. There's not one, except with single celled organisms, some of them. There's two."

So then where does the idea of gender enter in? Dr. Peterson explained. "What about personality? What about temperament? Well, by the social constructionist types, that's talked about as gender. But they're not very good with their terminology— extraordinarily imprecise... So sex is binary... temperament is not binary."

"So, temperament is gender?" I asked.

"Gender is a not good word," he responded. "It's a not good word because it's vague and it isn't measurable and it isn't definable and it isn't precise."

"Can we just say 'temperament'?" I asked. "What do we even need the word 'gender' for?"

"Well, I don't need it," he answered. "But what I would say is that people who talk about the diversity in gender are actually talking about diversity in personality and temperament, but they don't know it."

It sounds like we're splitting hairs. But the shift from using "temperament" to using "gender"—and especially with tying gender to identity—has had grave consequences.

"Ok, so now let's say you have a girl who's got a pretty masculine temperament," Dr. Peterson went on. "Well, does that mean she should transform her body? No! You certainly don't leap to that, especially given the catastrophe and the utter trauma and the mayhem of the hormonal transformations and the severity of the surgery. It's like, look, there are masculine girls. There are feminine boys. What are we going to do about that? Carve them up? Convince them that [they're the wrong gender] when they're three? That's not a good idea. It's certainly not the first thing you leap to, especially not when you also understand that... to some degree, adolescents are often extremely confused, not about their gender identity, you know, but about their *identity*." The error at the foundation of the idea of gender doesn't just cause intellectual confusion. It causes physical damage.

THE MEDICAL ESTABLISHMENT IS IN IT FOR THE MONEY

Knowing what we know now, I asked Newgent why she couldn't file a lawsuit against the doctor who not only completely botched her surgery, but also acted against such a preponderance of evidence in the scientific literature.

"Every single [attorney] turned me down," she told me. "Do you know why? Well, about the eighth one. I had an absolute breakdown on the phone with an attorney. I mean, I went nuts; somebody probably should have called the mental hospital. I was throwing—I broke a TV—I was throwing shit all around my apartment and I started to bawl."

She had been turned down so many times before, and she just wanted to know why. The case seemed like a slam-dunk. "Just will you please tell me why you're not taking my case?" Newgent asked the attorney. "And she told me that. 'Well, we looked at WPATH and there's no baseline to care. So to take your case, we have to create a baseline for care. That's millions of dollars. That little paper that you signed, you said that it's experimental.'"

In medicine, a baseline of care establishes common, minimal practices and conduct that assure patients are well taken care of and aren't harmed by negligence or abuse. When doctors fail to follow that baseline of care, they open themselves up to liability. However, because sex change surgeries are deemed experimental, baselines of care have never been established—and surgeons are rarely held accountable.

"What would be involved in creating a baseline of care?" I asked Newgent.

"It would take a lot of case studies," she told me—the type of case studies that don't exist for transitioning therapies and surgeries. "Companies like Lupron would actually have to run studies on hormone blockers to try to get it FDA-approved."

But if they submitted drugs like Lupron for approval, the entire world would learn the truth, not only about the terrible side effects of Lupron as it is currently used off-label for gender-affirming therapy, but also that transgender people who receive hormone therapy and who medically transition aren't actually happier than those who don't.

Lupron has been around for decades, as have sex change surgery and hormone therapy and transgenderism. The FDA hasn't approved drugs like Lupron for gender transitioning, and doctors haven't established baselines of care for sex change surgeries, not because we haven't had enough time, but because there hasn't been the will.

"Lupron refuses to do studies," Newgent told me, "because when they do studies, there's no doctor in the world that's going to sign up and go, 'Yep, I'm going to sign that, yep.'"

Some people really are true believers. They ignore contrary evidence, ignore the risks of experimental health care, and truly believe they are helping transgender people through medical transitioning. I believe Dr. Bowers, Dr. Forcier, and Gert Comfrey all fit in that category. But that doesn't explain everything. Something stronger and more pervasive than misplaced empathy is powering this medical machine. Like so many ills in society, these problems are exacerbated by money.

"Every child that they convince is transgender and in need of medical transition, it generates $1.3 million dollars to Pharma," Newgent told me.

I had no way to verify those numbers, but I could see how the costs could quickly add up. Every form of so-called bottom surgery from metoidioplasty to phalloplasty to vaginoplasty costs tens of thousands of dollars. Hormone therapy requires lifelong maintenance with lifelong costs. That's not to mention cosmetic surgery to make one's appearance seem more feminine or masculine. Then, after all that, there is the cost of responding to the side effects: brittle bones and heart failure and botched surgeries and the like.

"We have five children's hospitals in the United States telling girls that they can be boys at $70,000 a pop," Newgent told me. "You call them and say, 'Oh, you're doing phalloplasties on children.' 'Well, no. Well, only with parents' permission'—In a surgery that has a 67 percent complication rate, that will kill me from infection."

Other studies were more conservative—but not by much—on the risks. They confirmed high rates of complications with phalloplasty, including a 25 percent complication with the flap of skin used to form the phallus and a 64 percent complication with the extension of the urethra—the type of complication that left Newgent with debilitating pain and endless infections.

I asked Dr. Bowers earlier how much her surgeries cost, just to get a ballpark sense. "The cost of a vaginoplasty, for example, is—if somebody is paying out of pocket—it's around $30,000," she said, assuring me that this is "actually not overly expensive."

But then she told me, "We're finding that we have about 95 percent insurance coverage, thanks to many states now that recognize it as a real diagnosis and mandate coverage by employers of a certain size."

When insurance covers it, that's where the real money rolls in. "But what happens when you convince people that they're born in the wrong body? What happens to that prescription pad? What happens to that?" Newgent asked. "It becomes an illness that needs to be fixed. When it needs to be fixed, who has to pay for that prescription? Governments and insurance companies."

The timing lines up perfectly. The psychology DSM changed "gender identity disorder" to "gender dysphoria" in 2013. In seeming lockstep with that change, Medicare and Medicaid—the largest insurance providers in America—began coverage for sex change surgery in 2014. Private insurance followed suit, with coverage for sex change surgery spiking from around 25 percent in 2011 to 45 percent in 2014. All this opened up a floodgate of taxpayer-fueled funding for medical transitioning. By 2025, the sex reassignment surgery industry is set to reach a market value of more than $1.5 billion.

Is it any wonder that right at this time gender clinics began popping up everywhere? The resource "Trans Health Care" confirmed that "broader insurance coverage beginning in 2015... tipped the scales and kick-started the development of transgender surgery programs at medical centers across the country. There are now over thirty academic medical centers with sex change surgery programs."

These clinics and medical centers need customers though. It's no surprise that during the same period as all these insurance and clinical changes were happening, transgenderism splashed with full force on the culture scene, affirmative care was adopted that urged kids to adopt a gender identity different from their biological sex, and rapid onset gender dysphoria first made its appearance. All of a sudden there was a massive increase in the number of customers for medical transitioning. In the UK, the number of young people seeking "gender treatment" saw a 4,000 percent spike in the decade from 2009 to 2018. That's not a typo. It's actually 4,000 percent.

Dr. Grossman told me that during her twelve years at UCLA from 1996 to 2008, she saw maybe three or four transgender students. Today? "It's a different world now," she said.

Whether it's caused by the greed of big Pharma, the misplaced empathy of therapists, the false beliefs of doctors, the fear of parents, or the insistence of youth, the end result is the same: We are subjecting our children to dangerous experimentation that lines the pockets of the medical industry.

In Newgent's words, "We're taking kids that have suicidal ideation that already want to kill themselves. We're telling them there's something wrong with them. We're putting them in an experimental procedure... They still know who they are on the inside. They have to deal with the fact that they might look male, but they feel female on the inside because that's what they are. The truth is always there. So we're going to take those kids that don't fit in, all those kids that you really want to protect... then we're putting them into a part of society where they're going to be the most suicidal of any population in the world. And we're going to tell them they're going to be OK. No! They call me. That's why I don't sleep at night. It's why I'm on the phone constantly trying to find therapists for these people. This is why I look at media people and say, 'shame on you.' That's the truth."

Then she had a message to the parents who play along with this and who believe the lies they are being told at the expense of their own children: "You don't have the right to medically transition your child. You don't have that right. Nobody has that right but an adult, about their own body, with all the information put in front of them. I shame parents. Shame on you for doing that."

Behind the happy façade of "affirmation" and "identity" and "mental health" lurks a profound darkness. Proponents of transitioning speak with such gracious and welcoming words. They claim to have the best interest of their patients at heart. But they are wolves in sheep's clothing. They are butchers with a smile.

Transgender ideology is a heap of lies. The history is perverted. The science is falsified. At the very best we are inflicting irreversible damage on people for highly dubious ends. At worst, we are tearing people's bodies apart with no psychological benefit whatsoever.

The ideologues say this is a matter of autonomy and choice. Maybe so. Maybe people have the freedom to be unhappy and to cause themselves irreparable harm. But they aren't just hurting themselves. They are pushing it on our children. They are "affirming" every flight of fancy with medical interventions that no child can understand, and no person can ever take back.

They ignore the truth about everything—about biology and social science and DNA. But there is one truth above all that they cannot bring themselves to accept because it would tear everything they have built down to the ground. It's a truth so commonplace, that Newgent first learned it from her mother-in-law. "I have a mother-in-law who is fruity as a fruitcake, but I love her to death. And she always tells me that happiness is an inside job. I always tell her to f*ck off. But she's right. Because at some point you have got to go 'Aha, it's an inside job.'"

That's the twist of transgenderism. They make you believe happiness comes from affirming who you really are on the inside, but then they tell you that you can only be happy if you manipulate your body from without. They tell you that you have to accept who you are, but then they sell you an endless and expensive bill of goods all so that you can change who you are.

Lies can never stand on their own. That is why when people begin to doubt, the defenders of the lies resort to force. Gender theory first spread through education, pop culture, and the soft reach of social media. Yet there is still resistance. The next phase of the assault won't give up on the old weapons. They will just add a new one: coercion.

CHAPTER 8

THE TRANSGENDER CULTURE WAR

WHEN I SET OUT to discover what is a woman, I didn't think I would stumble on a cabal of insurance providers, medical personnel, leftwing culture warriors, educators, academics, therapists, journalists, sexologists, pop culture icons, and so many more all operating in lockstep to feed adults and children alike into the expensive and irreversible Frankenstein experiments of gender transitioning.

In all of this searching and discovery, nobody had actually been able to tell me yet what is a woman. But what I had learned is that, ironically, the people who clearly have no idea how to answer that question also have the most power over what so many millions in our society believe and do when it comes to sex and gender. They are funneling people through experimental therapies and surgeries with no measurable benefit. They have made innumerable people believe obvious falsehoods like the idea that a woman can have a penis. And their power is only growing.

Trans ideologues have long tried to win the debate with persuasion and propaganda. We still have much of that—from transgender reality star Jazz Jennings promoting his alternative lifestyle on TLC to the now-infamous ad from Gillette warmly depicting a young woman so pumped through with hormones that she now needs to shave her face. As a result, the transgender cause has made impressive headway.

Transgenderism has gone from unknown to accepted to celebrated in an historical blink of an eye. The number of trans youth has skyrocketed. Clinics are opening everywhere. You can't turn on the TV or read the paper without being confronted with something, anything, that's queer. By all accounts, they are making converts every day.

You'd think that would be enough for them. After all, most people I know—and honestly most normal, well-adjusted people in America—really don't care that much about the opinions of people we've never met. By and large, we just want to live our lives, go to church, make a good living, and enjoy the sweet company of family and friends. Sure, we don't like it when others make bad decisions. Yes, we want a modicum of decency, cleanliness, safety, and order in our communities and in our nation. We obviously don't prefer it when people hate us. But, as long as they let us raise our kids and live in peace—and as long as they keep their crime and filth and perversions far, far away from our children—we're pretty fine with a live and let live approach. Of course, I'm talking in an ideal world here. Private decisions always influence the public sphere. But the broader point is that we don't demand that everyone think and act and live exactly like we do. To put it in leftwing terms, we really are fine with a bit of diversity.

That's what makes it so hard to understand the trans activists out there who demand that we absolutely *must* accept their ideology, wholly and entirely. They can't tolerate that the more we learn about transgenderism, the less we like it, and that we won't sacrifice our children to their medical machine. There is a core of dissidents who have seen the truth and won't turn their back on it. And every time we refuse to say "menstruating people," every time we continue to just say "man" instead of "trans woman," every time we refuse to change someone's pronouns just because they tell us to, it reminds them that we see behind the façade and we know exactly the damage that transgender ideology has done.

But if we're being honest with ourselves, we know that even if we did change our language and follow their ever-altering rules of speech, it wouldn't be enough. Of course, they want us to say it. But they also want us to mean it in our hearts.

As I learned before, the very lifeblood of transgenderism is affirmation. According to them, affirmation is the key to everything—to better grades, better jobs, better habits, a better life, and, above all, better mental health. Everything in their lives depends upon affirmation, and affirmation comes from others. They cannot be happy alone. They cannot find peace from within.

That means when you fail to give affirmation, you not only undermine a transgender person's happiness, you are a danger to his or her "mental health." And it's not just that one person you harm; it's *all* trans people. When you reject one person's gender identity, you reject the idea that their identity is rooted in gender at all. That makes you more than a menace to one person's mental health. It makes you a threat to everyone's public health as well. And if we've learned anything over the course of the coronavirus pandemic, it's that threats to public health cannot and will not be tolerated.

That's why the trans movement won't stop with persuasion and propaganda. They feel entitled to use force. Trans ideologues have every intention to use the powers of the state to crush dissent because to them dissent is illegitimate and dangerous. It's only a matter of time before it comes to that (and I'll explain how they plan to do it in the next chapter). But before hard totalitarianism is enforced by the law, gender activists are using soft totalitarianism to crush dissent through social and economic pressure. A poisonous cocktail of social influence, libelous public campaigns, alteration of language, and the power of private industry is being leveraged to impose pain on anyone who resists their agenda.

ALWAYS THE VICTIM, NEVER THE VICTOR

Part of the reason the transgender activist machine won't stop is that, despite all the evidence, they don't actually think they are winning. Until everyone submits, there is more work to be done. Nobody put this better than Dr. Patrick Grzanka, the chair of the Women, Gender, and Sexuality Interdisciplinary program at the University of Tennessee, Knoxville. It was a beautiful day when I arrived to speak with him at his office in the University of Tennessee's flagship campus: sunny, perfect weather, and the campus was abuzz with the next day's big football game against the Georgia Bulldogs. My reception inside Dr. Grzanka's office was much less warm than the feelings outside.

Dr. Grzanka styles himself as a "scholar-activist," which, in modern academia, is also known as redundant. Right off the bat, he rejected the idea that society has embraced transgenderism.

"I don't know that we have a lot of evidence today to say that our society is more accepting," he informed me. "We might see greater visibility of gender diversity, but we also see retrenchment around gender diversity in virtually all aspects of social life. The entire country has been awash in anti-transgender bills in schools prohibiting participation—like in my own state, the state of Tennessee—participation of trans athletes in women's sports. We've seen bathroom bills all across the country that try to police the way that trans people use restrooms in public spaces."

He expanded that he believes those bills reveal "at the very least institutional transphobia to the extent that we see structures like the law, education, and sports continuing to use reductionist and transphobic ways of thinking about gender identity and gender expression," soon adding that he thinks this "retrenchment" in the face of the "diversification of gender identities" is mostly occurring in "historically conservative institutions like sports and education."

I wasn't sure what planet he was on. As far as I could tell, sports is where professional athletes all kneel for the national anthem, and millionaire superstars complain about how racist everyone is. As I already discovered, education is where the most radical forms of sex education and gender identity are force-fed to children. I've never met anybody who saw those as conservative institutions—at least not recently.

Regardless, it was clear from our conversation that Dr. Grzanka sees a lot more work to be done to get the American public to accept transgenderism.

He mentioned bathroom bills, so I decided to explore this issue a little more. That's when I saw Dr. Grzanka use one of the most common and most sinister tactics that gender activists use to shut down debate and get you to submit. It all boils down to viciously insinuating that you are an evil person who is trying to hurt, if not kill, the weak.

He set off his rhetorical bomb after I asked a simple question: "What about the women who say, 'Listen, I don't feel comfortable in a restroom or a locker room with a male body.'" I thought it was a rather legitimate question posed by, well, rather legitimate women.

Dr. Grzanka rejected what I said in principle. "The way you phrase the question, you know, it betrays the idea that you don't believe that trans women are women. So in that case, you know, I don't really know how to even start a conversation or debate about that because you are *denying someone's existence* as the sort of entry point." (Emphasis added)

The accusation was understated, but immensely threatening. He didn't accuse me of being wrong. He didn't say it was a bad question to ask. He didn't refute my point in any way at all. He accused me and my words of doing some sort of existential violence against people. There is no response to that. It is a statement meant to shut down debate.

The fact that I actually do believe that transgender people exist wasn't the point. The implication was that a person's gender identity is so wrapped up with who they are—with their very essence as a person—that to disagree with that identity is to erase them from life itself. It elevates gender to a higher plane, making gender a first principle upon which everything else depends. This is why there is no limit to what gender ideologues will do to get us to submit. To them, this is a battle about life and death.

Such underhanded accusations might stop some people, but it didn't stop me. If "trans women" and women are the same, I asked him, then what are women in truth and reality?

"You keep invoking the word 'truth,' which is condescending and rude,'" he spat at me. "I'm really uncomfortable with that language of 'getting to the truth'... because it sounds actually deeply transphobic to me."

There he goes again with the ad hominem attacks. I ask a simple question; he accuses me of transphobia. The goal is not to get me to accept transgenderism; it's to force me to submit out of fear of being labeled a hater.

"How is the word 'truth' condescending and rude?" I asked him calmly.

"You're walking on thirty seconds more of thin ice before I get up," he said, threatening to shut down the interview on the spot.

I tried to keep him calm. Obviously, he wasn't used to people questioning his ideology, even in an extremely mild way. I guess being a professor at an "institute of higher learning," he wasn't used to debate. He was hoping to have us produce a puff peace, not conduct an actual interview.

I kept him calm enough to keep going a few minutes more. I told him I'm not there to lead him on. I am truly trying to understand what he believes. He is a professor of women, gender, and sexuality studies, so theoretically he should know how to define the word "women," after all.

"I don't go around talking about capital 'T' Truth," he told me, "because it is often used as a weapon against people whose ideas, experiences, and lives don't fit so neatly into the kinds of stories that oftentimes books with bad intentions try to tell."

It was clear from his tone that he was accusing me of having bad intentions. It was also clear that he was very, very afraid of the idea of truth. I had learned a lot about the truth so far in my journey, and for a radical like Dr. Grzanka, his fear of the truth was very understandable. After all, the truth would tear his entire world down.

I tried asking him why he was avoiding the question. Why couldn't he just say what a woman is? Finally, the interview completely broke down. He turned to the side and looked at one of the crew with me. "Could you keep your mask on in the building, please," he shouted. We had been talking to him for over a half hour without masks on, and he had expressed no issue whatsoever up to that point. It was clear he was using masking as a power play to get us to bend to his will in some way since we refused to blindly follow his proclamations on gender ideology.

We got an email after the interview ended threatening legal action against us—honestly, another common tactic the pro-trans left uses to silence debate and quash the truth.

THE TRANSGENDER WAR ON WOMEN

Dr. Grzanka may not have wanted to consider how biological women are threatened by the imperialistic advances of transgenderism. But there was someone I spoke with who had to consider it because she faced it personally year after demoralizing year. Her name is Selina Soule, and she's a track athlete who was forced to compete against biological males throughout high school. Now, she still runs track as a sophomore in college.

As we stood out on a cold track loop during a blustery fall day, I asked her what happened.

"Let's go back to high school," I said. "You're on the track and field team for..."

"Glastonbury High School," she responded—a normal, run-of-the-mill high school in Connecticut outside of Hartford.

"What exactly happened?" I asked.

"So, throughout all four years of high school, I was forced to compete against biological males. I only competed against them in the sprinting events. But I raced against these athletes over a dozen times throughout the years, and every single time I lost. I was never close to winning."

It all happened because Connecticut changed its rules governing women's sports. The only thing that biological males had to do was sign a piece of paper informing the school that they changed their gender and that was that!

After that policy change, two boys changed their gender and competed against high school girls. According to Soule, the girls didn't have any hope.

"When I got to high school, my goal was just to get good grades and pave my way to college and compete on a college track team," she said. "I never thought I'd have to face this issue."

Every single race the boys were in, it wasn't close. Soule continued: "They beat me, by 20 meters, out of qualifying spots. I missed out on qualifying for New England's. I had to go in the long jump and 4x200 meter relay"—not her usual events—"so I was forced on the sidelines in my own event. If they were not there, I would have been able to qualify. I missed out on so much throughout my high school career."

"Did they win all the events or almost all the events?" I asked.

"Between the two of them, they won every single event they competed in," she answered.

It wasn't exactly a great mystery why they won.

"It has been scientifically proven that men have great physical advantage over women," she informed me. "Everything from muscle mass to skeletal structure, organ size; of course, testosterone is probably one of the most notable ones. So there's everything together that creates such a big difference between a man and a woman's body."

Soule wasn't the only one to experience this difference. Recently, a biological male named Lia Thomas joined the women's swim team at the University of Pennsylvania. He quickly became the number one college "woman" swimmer in the country. After one of his wins, he taunted the biological women he was up against, saying, "That was so easy, I was cruising."[1] Yeah, hard to imagine why it was easy. Of course, the actual women Thomas was racing against were completely discouraged because they knew going in that they were going to lose.

I asked Soule if these biological males needed to undergo any sort of medical intervention—perhaps cross-hormone therapy—before they could compete against females.

It seemed unlikely. "One competed as a boy for three seasons and then in the two weeks between the indoor and outdoor season, transitioned to competing on the girls' team," she told me. Two weeks is hardly enough time for hormone therapy to have a significant impact. Not to mention it doesn't matter. All the pro-trans activists kept telling me that physicality had nothing to do with one's gender identity. I'm sure they would say that requiring biological men to take estrogen before they compete in women's sports (ignoring, for a brief moment, the incredible physical harm that causes) would also be "transphobic."

When it comes to transgender people competing in women's sports, the concern always seems to focus on how the transgender person would feel if he isn't allowed to compete. Few people ever ask how the women athletes feel having to compete against biological men.

"After so many losses, it just gets to the point of why am I even doing this? Why do I keep training so hard and sacrificing so much just to place third and beyond?" Soule reflected. "Going into races knowing that we will never be able to win is just so demoralizing. We shouldn't be competing for third place and beyond. We should be competing for first place."

Soule's feelings aren't a bug of the transgender cause but a feature. They want to break our spirits so that we don't resist what they are trying to do. And she had no way out. It's not like she could set up another league for biological women only. Biological men would then jump straight into that in the name of transgender rights.

But Soule didn't go down without a fight. Not only did she speak out against the injustice being done to her and to every girl competing in high school track, she filed a Title IX complaint alleging sex discrimination. When the federal government dragged its feet, she filed a lawsuit against the Connecticut sports governing body.

I asked her if other girls on the track field agreed with her.

"None of the girls that I competed against thought it was fair," she said.

"But none of them said anything?" I asked.

"At the beginning, I was the only one that was actually willing to speak out and try to get this policy resolved," she said.

There was a simple rationale for this. Everyone else was afraid. They knew they couldn't just disagree on this issue and let bygones be bygones. They knew that the pro-trans side would come for them, threaten them, and try to destroy their lives. "They were all too afraid of speaking up," she told me, "because of retaliation from the media and from coaches and administrators and everything... They were afraid that it might affect their college recruitment, and that nobody would want them if they got involved in this issue."

I asked her if those fears were justified.

"I got called names and received death threats," Soule informed me with a striking calmness for a nineteen-year-old young woman. "I did get a lot of hate from the media."

She also received a lot of anonymous comments online and on social media. "Most are saying that I'm a sore loser, and I should just try harder if I want to win," she said. "There's nothing that I can do to compensate for the great physical advantages that those athletes have over me and my fellow female competitors."

All this was done to a high school student in her teens. She wasn't committing acts of violence or selling drugs. She wasn't marching with the KKK or starting a chapter of Hitler Youth at her school. All she wanted was to be able to play her sport and compete fairly. For that, the outrage mob tried to break her down. They attacked her viciously and personally. They wanted to make her an example to show that some opinions and some actions are unacceptable and will not be tolerated.

How did she hold up through it all? "I was more focused on restoring fairness to track and field," she said, "so no other little girl would have to experience the pain and heartbreak I had to go through my four years of high school."

Selina Soule is a brave young woman, braver than most grown men I know.

THE TRANS CULTURE CRUSADE

Personal attacks and retaliation have become par for the course for the LGBTQ mafia. They know that people really aren't free to say and act the way they want if doing so costs them dearly. Selina is far from the only victim.

Former pitcher and now former ESPN analyst Curt Schilling learned this the hard way as well after saying something that just about everyone agreed with a few years ago. On Facebook, he posted that "a man is a man no matter what they call themselves [sic]. I don't care what they [sic] are, who they [sic] sleep with, men's room was designed for the penis, women's not so much. Now you need laws telling us differently? Pathetic."

Despite taking down the post and submissively issuing an apology, he was still terminated from his job. ESPN announced that he wasn't welcome there because it is an "inclusive company," which, of course, cannot include anyone who outwardly believes that only men have penises.[2]

J.K. Rowling, the author of the Harry Potter series and once a feminist icon, also learned that you can accept 99 percent of the left's sexual and social dogma, but if you deviate on even one point, you are anathema. It all started, like so many tempests in a teapot, with a tweet. Rowling wrote: "'People who menstruate.' I'm sure there used to be a word for those people. Someone help me out. Wumben? Wimpund? Woomud?"

The pro-trans left didn't find her humor funny, but unlike Schilling, Rowling didn't back down. In fact, she defended her views based on her own leftwing principles, proving that while logical consistency may be a dying breed on the left, it is not yet extinct.

"If sex isn't real, there's no same-sex attraction," Rowling wrote. "If sex isn't real, the lived reality of women globally is erased. I know and love trans people, but erasing the concept of sex removes the ability of many to meaningfully discuss their lives. It isn't hate to speak the truth." She continued, "I respect every trans person's right to live any way that feels authentic and comfortable to them. I'd march with you if you were discriminated against on the basis of being trans. At the same time, my life has been shaped by being female. I do not believe it's hateful to say so."

Her defense was insufficient. Rowling was labeled a "TERF," a term of derision used on the left that means "trans-exclusionary radical feminist." Echoing how people spoke of the evil Voldemort in Rowling's Harry Potter series, fans of the author labeled Rowling "she-who-must-not-be-named." Fansites kept up devotions to her work, but removed photos of her, links to her website, and now avoid even naming the author whenever possible.[3] Two actors who have built their entire careers off of their roles in the Harry Potter films—Daniel Radcliffe and Emma Watson—put out public statements condemning Rowling. For confirming the reality of biological sex, Rowling was labeled by the media as "anti-trans."[4]

Thought policing is exceedingly prevalent in the intellectual world as well. Abigail Shrier—author of the book *Irreversible Damage,* which documents the rise of transgenderism and the impacts of medical transitioning, especially on children—faced a coordinated campaign of public erasure. Her book barely made it to print.

Her first publisher backed out of the book deal after staff at the publishing company complained.[5] After Shrier found a second publisher, Amazon banned that publisher from buying ads to promote the book, even as Amazon allowed sponsored ads for books that promote medical transitioning for teenagers.[6] Book reviewers ignored the existence of her work, and when Shrier went on the Joe Rogan podcast to promote her book, employees at Spotify—the company that hosts the Joe Rogan show—threatened a strike and called for the episode to be removed from the internet.[7] A professor at UC Berkeley, the University of California system's most prominent and prestigious school, called for the book to be burned. Target pulled books from the shelves after complaints on social media, only returning it to stores after proponents of free speech (who had finally woken up to the aggressive censorship campaign) hammered the retail giant for its cowardice.

The censorship campaign didn't end after her book came out. Her attempt to speak at Ivy League Princeton University in December 2021 generated such a large degree of threats and harassment that organizers had to hold the event off campus in a secret location.[8] As if all this wasn't enough, another chapter in this story demonstrated the lengths that the gender activists will go to in order to shut down debate and hide the truth. A local parent wanted to raise money on the crowdfunding site GoFundMe to put up a billboard promoting Shrier's book. Proving that the left deems no fish too small to fry, GoFundMe shut down the parent's fundraising page. At the same time, GoFundMe hosted tens of thousands of fundraising pages for people to raise money to get "top surgery" and "bottom surgery."[9]

Shrier's work drew from the same research study on rapid onset gender dysphoria (ROGD) by Dr. Lisa Littman that I discussed earlier. After Littman's finding that transgenderism spread among adolescents like a social contagion gained notoriety with the general public, trans activists balked and the publisher of her research demanded Dr. Littman "correct" her research.

To Dr. Littman's credit, while she did add information as well as "more detailed descriptions about recruitment" of those who were in the study, she declared that "the results section is unchanged in the updated version of this article."[10]

Even Scott Newgent, the transgender person I spoke with who relayed the grueling details of her own transition and the ways the medical industry targets children, has faced censorship. Twitter proclaimed a lifetime ban on Scott for posting a video showing actual pictures of medical transitioning that you would see in a medical textbook along with the facts that girls as young as thirteen are getting mastectomies and that puberty blockers cause early onset osteoporosis. The information isn't debatable. It's established fact. Yet these facts contradict the trans movement's ideology, so they are banned.

Revealingly, posting the same medical images on social media while affirming gender transitioning faces no censorship.[11] Perhaps most ironically, so many of these people being censored and attacked from Selina Soule to J.K. Rowling to Abigail Shrier to Dr. Lisa Littman to Scott Newgent are all biological women, a group that was for no short period of time deemed a special protected class by the very same people on the front lines of the transgender crusade.

The erasure of wrongthink from the public consciousness is a common tactic of the left now being used by transgender activists to shut down disagreement. People cannot think what they have never heard of, after all, so if you don't let them hear about alternatives to transgender ideology, then more people will accept it.

It's this same rationale that led an organization like Netflix to quickly change the name of actress Ellen Page to "Elliot Page" on all of her earlier movies as soon as she came out as transgender.[12] That Ellen/Elliot was portrayed as a woman and outwardly identified as a woman at the time earlier movies like *Juno*, *Inception*, and *X-Men* came out didn't matter. In a very Soviet way, the past was changed to fit her new identity.

This ties into the transgender concept of "deadnaming," where someone uses the name that a transgender person's parents gave him or her at birth instead of the name that person has chosen for him or herself after transitioning. Like accusations that rejecting transgender theory "denies the existence" of trans people, claiming someone has "deadnamed" a trans person is just another weapon to scare people into submission by accusing them of a form of murder. Deadnaming doesn't even have to be intentional. If you accidentally called Ms. Page "Ellen" instead of "Elliot," you have effectively killed her transgender self by failing to recognize her new identity. Public recognition of the lie is mandatory.

Enforced public acceptance is hardly limited to names, books, or movie sites. The transgender culture war has extended far into the public square as well, demanding conformity and affirmation in every institution. Our taxpayer dollars are now being used to fund transgender surgeries, not just through Medicare and Medicaid, but in the military as well. But don't worry; the military has assured us that transgender persons will be held to the same medical fitness, physical fitness, and deployability standards as everyone else.[13] People get rejected from joining the military for having eczema or because they broke a bone when they were kids, calling those pre-existing conditions. How a post-op trans "woman" who takes regular, physically harmful hormones and must do continual maintenance to "keep the bathroom walls open" can remain deployable is beyond me. Obviously, this change makes our military weaker. But that's not the point. The triumph of trans is the point.

Schools are another active theatre in their culture war offensive. The pro-trans side knows that if they target children when they're young, malleable, and when their parents aren't watching, then they will reap dividends years down the line. The kids don't need to transition to become "allies" because the full breadth of transgender ideology is being taught to them as an unquestioned first principle from the youngest age.

I've already documented just how deep transgenderism has inserted itself into the education system. But this is about much more than putting books in libraries and classrooms. Teachers and administrators are going to extremely great lengths to brainwash children and purge dissent.

Leaked audio from the California Teachers Association (CTA) revealed teachers plotting how to recruit students into LGBTQ clubs without their parents finding out. They don't officially form the club or keep rosters, all so they never have to reveal who shows up. "In fact, sometimes we don't really want to keep records because if parents get upset that their kids are coming, we're like, 'Yeah, I don't know. Maybe they came?' You know, we would never want a kid to get in trouble for attending if their parents are upset," one teacher was recorded saying.

This is a revolution driven from above. Oftentimes, students don't actually want to spend all their spare time talking about their gender and sexuality... surprise, surprise. That's when teachers started stalking—to use their own words—their students online in order to recruit them into their faculty-formed LGBTQ clubs. "We got to see some kids in person at the end of last year, not many, but a few," one teacher said at the end of the 2020 school year. "So we started to try and identify kids. When we were doing our virtual learning, we totally stalked what they were doing on Google when they weren't doing schoolwork. One of them was googling 'trans day of visibility,' and we're like, 'check. We're going to invite that kid when we get back on campus."

The teachers also track student conversations amongst themselves to see who is most committed to their ideology. "Whenever they follow Google Doodle links or whatever, right, we make note of those kids and the things that they bring up with each other in chats or email or whatever," continued the teacher. "We use our observation of kids in the classroom—conversations that we hear—to personally invite students, because that's really the way we kinda get bodies in the door."

Another recorded teacher spoke of how she purposefully volunteered to do morning announcements at her school in order to regularly include LGBTQ propaganda. Another spoke about doing a "little mind-trick on our sixth graders" by presenting gender materials as soon as possible so that by the time their parents get word of it, they won't be able to stop it.

What were the teachers' responses when parents complain? "Thank you, CTA, but I have tenure! You can't fire me for running a GSA. And so, you can be mad, but you can't fire me for it."[14] They call it equity. I call it the legally protected corruption of our children.

Lest you think this is only happening in crazy deep-blue California, laws everywhere from Wisconsin[15] to Florida[16] require staff and teachers to hide a student's gender identity from his or her parents. Teachers in Maryland have even been coached to switch between using a student's preferred name and pronouns at school and then to use the student's given name and biological pronouns whenever his or her parents are present.[17]

The school system isn't just targeting children, though. Teachers who fail to toe the line face such a hostile work environment that many choose to leave rather than become complicit in the indoctrination. After Loudon County, Virginia, adopted a policy to allow biological males to use the girls' locker rooms and restrooms while requiring all teachers to use the preferred name and pronoun of students, one teacher was forced to leave her job. She tendered her resignation publicly and directly to the school board. "This summer, I have struggled with the idea of returning to school knowing that I'll be working yet again with a school division that, despite its shiny tech and flashy salary, promotes political ideologies that do not square with who I am as a believer in Christ," the teacher said. "Clearly, you've made your point. You no longer value me or many other teachers you've employed in this county." Another teacher in the county was suspended just for speaking against the policy at a school board meeting.[18]

A similar controversy bubbled up in Indiana where a teacher was forced to resign when he refused to use students' preferred names and pronouns. The teacher even tried to meet the school and students in the middle by only using last names. But this was obviously never going to work. Coexistence and accommodation are contrary to the real goal of transgender activists, who desire nothing more than uniformity.

That teacher's case went all the way up to a federal judge, who ruled against him, effectively saying that his religious beliefs didn't matter and that his refusal to deny biological reality created "undue hardship on students." Mimicking the language of the trans ideologues, Judge Jane Magnus-Stinson wrote that the teacher's "opposition to transgenderism is directly at odds with [the school district's] policy of respecting transgender students, which is grounded in supporting and affirming those students."[19] Got that? Opposing transgender ideology is enough to get you fired and to have the court system rule against you.

INDECENT EXPOSURE

There are no lengths that the left won't go to in order to force people like you and me to embrace transgenderism. They will even force women to look at men's penises and then berate them as haters and bigots if they express any discontent. Think I'm joking? This actually happened at Wi Spa, a Korean spa in Los

Angeles. After women—including a six-year-old girl—complained about the presence of a biological male in the women's section exposing himself to them, the staff told them that they couldn't do anything because the man identified as a woman. The man himself later claimed that he was a victim of transgender harassment.

I asked Gert Comfrey, the kind, non-binary therapist I had been talking to, about what should happen in these circumstances. Her response revealed, wittingly or not, that behind the pleasant façade of affirmation and love and acceptance rests a deep-seated determination to accept harassment in defense of an indefensible ideology.

"To say, 'oh, I don't want, like, a penis in a women's locker room is trans exclusionary because some women have penises," she told me. "I think this also plays into this narrative that comes up a lot that trans people are predatory... and, in fact, trans people are the people who are being assaulted. Trans people are the people who are being harassed." But I asked her about the documented fact of women who don't feel comfortable or safe having a naked man in front of them in a locker room or a spa.

She effectively said, tough luck. "There's no male involved here, and well, maybe there is. I'm not sure if we have a cisgender man walking into, like, a women's room or walking into a women's locker room and is attempting to be, like, predatory or exhibitionist. Like, that's a problem," she told me. "Ok, so we need to invite that person to leave. But if what we're talking about is a transgender woman wanting to access a women's locker room... that's our work as a society to respect that."

I can't imagine such a man being "invited" to leave and accepting the invitation. But in the Wi Spa incident, it really didn't matter, seeing as the man proclaimed he is transgender and, as such, we are then required to "respect that." I guess the fact that the man in question apparently walked up to the women's hot tub with an "entitled attitude" and a bearded face, according to one witness, then sat down at the edge of the hot tub with "his genitals fully on display" just doesn't matter. Biological women have no right to private space any more.

Hidden in plain sight in this whole controversy is the fact that this is direct sexual harassment of a young child by a grown adult male. But the left are not only ok with this, they celebrate it. That man is living his truth—well, "her" truth to them—and there's nothing you can do about it. What does it matter that he has a penis and a beard? It's not like he even has to make any attempt to look female. All he has to do is proclaim he is a woman, and then you have to accept that.

The ugly truth is that we're easily heading down a slippery slope where we normalize this kind of degeneracy and evil. Of course, the fact that a six-year-old was there didn't stop the man. It may have even been a reason he did it in the first place. Like Alfred Kinsey and John Money and explicit sex and gender education for elementary school students, this is clearly another attempt to normalize the sexualization of children and to say there is nothing wrong with it. It's just another step to ultimately promote pedophilia, though the left would never use that word. That's where this is ultimately going. The left may deny it right now because that word has a bad association, and it can never be rescued. But they will start calling it something else.

In fact, they already have. The term is "minor-attracted persons." Notice the Money-esque bait and switch, changing it from the action or proclivity of "pedophilia" to the *identity* of "minor-attracted persons." Once something is claimed as an identity, it can't be rejected or argued with. You deny someone's existence to express your disagreement, after all.

The normalization of the concept of "minor-attracted persons" is already well at work in academia, where all these evil ideas started in the first place. A 2012 paper from the University of British Columbia in Canada began building the links, modifying the Kinsey sexuality scale to include "minor-attracted persons."[20] A professor at Old Dominion University in Virginia argued that we really should stop using "pedophile" and start using "minor-attracted person" because it is less stigmatizing.[21]

These are small rumblings at the moment. But if we've learned anything, it's that radical ideology at universities pushed by unrepentant perverts has a way of going mainstream fast.

The Wi Spa incident led to protests outside the spa where Antifa showed up, and the media claimed that anyone trying to defend innocent women and children from having a penis shoved in their faces at a spa were far-right extremists. They claimed that by shouting "save our children," protestors were effectively QAnon conspiracy theorists. Left largely unsaid is the fact that the perpetrator was prosecuted on six counts of indecent exposure back in 2018 as well.[22]

THE BRAVE CASE OF DON SUCHER

Obviously, protecting children and not wanting men to disrobe in front of women isn't a right-wing position. But the left knows that if they politicize it and defame anyone who resists, fewer and fewer people will be brave enough to complain.

But I did speak with one person who was brave enough to resist the outrage mob. His name is Don Sucher, and he was targeted and filmed by a transgender person in an attempt to take him down.

Sucher is a man of profound common sense. He was offended at the insanity of calling biological men women and biological women men, so he put a sign up in his *Star Wars* memorabilia store in Aberdeen, Washington, that simply said, "If you are born with a d*ck, you are NOT a chick." The fact that he felt compelled to post this at all and that it would be controversial is indicative of our times.

Soon after, the short but sprightly Sucher was confronted by a ponderous, so-called trans woman on the city council. I asked Sucher what happened.

"He came around the corner, and I thought, 'Ok, I recognize him,'" Sucher told me as we sat in his store, every nook and cranny packed with *Star Wars* figurines and trinkets. "I says, 'Oh, I recognize you. You're our new city councilman.' He says, 'No, I'm a city council-woman.' And that flipped a switch."

The entire time, the councilman had someone filming the inter-action, obviously intending to post the video online so that Sucher would be attacked for his "anti-trans hate."

I asked him if he thought people are confused in our society today.

"I don't think they're confused," he said. "I think they've just lost all sense of reality."

Does that bother him at all?

"If that's what you're identifying as, fine. Just come in, have fun, buy something, leave, come back, buy some more. I do not care. Man, it's so simple."

Sucher seemed like a straight shooter, so I asked him how he thought all of this started—and he hit the nail on the head. "Well, supposedly, I think this has been going on for years. Slowly. It's like a cancer started years, years, and years ago. Now it's feath-ering into the schools. Now it's feathering everywhere. And we give them an inch, and they take a mile. And they've got the news media behind them... I just don't understand how a normal human being could believe that sh*t."

I asked him why so many others are so afraid while he doesn't seem to care at all about the backlash.

"Feelings?!" He said incredulously. "I don't give a sh*t about their feelings. I'm old, and I give up [on] 'feelings' bullsh*it. These people... they're just insane. Again, you can act like you want to be an ocean wave, if you want, but don't come into my store and tell me that's what I got to believe."

I guess being old does have its benefits. "A lot of people are terrified to even talk about the subject," I stated, and Sucher can understand why.

"Right, and why?" he interjected. "It's because I think they are scared. They're afraid of their job. They're afraid of the repercussions. They're afraid something will happen to their family. And I understand that, you know... I had a guy from England write me a note and said, 'You know, man, if we had done what you'd have done in England,' he said, 'We'd have been in the clink.' Wow."

Blessedly, no American that I could think of has yet been thrown in prison for failing to affirm transgender ideology. But that's not because the trans ideologues don't want to. Right now, their cultural influence is enormous. They can get people fired, generate death threats against their enemies, push people out of their jobs, box parents out of the lives of their children, and publicly defame anyone who stands in their way, from internationally renowned authors to no-name shop owners. Despite all that, it's not enough. They don't just want to make dissent unacceptable. They want to make it illegal. And violators will be prosecuted to the fullest extent of the law.

CHAPTER 9

CRUSHED BY THE BOOT OF THE PINK POLICE STATE

ROB HOOGLAND WAS sitting in a jail cell.[1] In some ways it was to be expected. He did break the law, after all. But that didn't make it right.

He didn't know when he'd be released. He was in uncharted territory. Few people had been prosecuted for the reason he had. But as he sat there, I'm sure he had a lot of time to think about how it all happened—how a man like him, a father, a postman, an average Joe, ended up in jail, treated like a gang member, robber, rapist, or murderer. He was just trying to save his daughter. But what he did made him a criminal.

It all started back in 2015. His then fifth-grade daughter—let's call her "Mary"—was getting into trouble at school. The divorce between Rob and his wife that year was obviously hitting young Mary hard. Rob and his ex-wife were probably not on good terms, but they did agree that it would be in Mary's best interest to see her school counselor. It seemed to be helping, or at least it wasn't hurting. Mary continued to see the counselor into seventh grade when, all of a sudden, things started to change.

Mary came home one day with her hair cut very short. Rob soon found her yearbook. Mary's name wasn't there, but her picture was. Under her picture was a new name, a masculine name. By the time Rob figured out what was happening, Mary was already well into the process of socially transitioning. A psychologist had encouraged her to take cross-sex hormones, and she was already referred to a doctor who would give her the prescription.

Rob attempted to retake control, but the totalizing transgender ideology had already sunk its roots deep into Mary's psyche. "If you don't let me take testosterone, I'm going to kill myself," Mary told her father. But Rob knew her better than that. "No, you know you're not," he told her. I can imagine Mary hesitating, realizing that the line didn't work as intended. "I know, but they told me to say that," she responded. She was kept from taking cross-sex hormones for a time. But her social transition continued.

By 8th grade her school made a special bathroom arrangement for her. Everyone was required to refer to her by her new, chosen name. She saw a top transgender psychologist in the region. Rob asked the psychologist to treat his daughter for depression, seeing all the telltale signs there. The psychologist refused, saying that cross-sex hormones would "solve all her problems." She was sent to an endocrinologist, a hormone therapist, and after a quick hour consultation, Mary was given puberty blockers and prescribed testosterone. At the time, she was only thirteen years old.

Soon, Mary was brought by her mother to the hospital to receive testosterone via injection. As a minor under custody of both her mother and father, Mary needed both parents to grant their consent. Rob said "no." His wife may have bought into the lie. His daughter may be suffering under the lie. But he would not sacrifice his daughter to it.

He didn't know it at the time, but that "no" set Rob on the path to prison.

The endocrinologist determined that Rob must be blocked out of his daughter's medical decisions and prevented him from having access to her medical records. The hormones would be given without his consent. But Rob's "no" was more than a rhetorical act. It was a statement of principle. He would not go down without a fight for his daughter. So, he appealed the decision to the courts.

Though Mary was only fourteen years old at the time, the judge ruled that Mary was mature enough to make medical decisions for herself. She didn't need her parents' approval, and her own consent was "sufficient for the treatment to proceed." Rob must have been crushed at the decision. But that's not all the judge decreed.

Rob was summarily banned from trying to persuade his daughter to stop her so-called treatments. He was banned from addressing her by the name he and his ex-wife gave Mary when she was born. He was banned from referring to Mary as a girl or using feminine pronouns to describe her in any conversation with anyone at any time. Failure to follow these rules would be "considered to be family violence," the judge said.

Rob had a choice. To even speak the truth would be considered a crime by the state. But to fail to speak the truth would violate a higher law, a law that governs every father from the very moment the spark of life animates the soul of his child. Rob had a duty to fight for his daughter. No law could relieve him of that duty. No law could stop him.

Rob appealed the case. Furthermore, he resisted the judge's order and spoke in public about what was being done to his daughter. Another judge stepped in to stop the publicity. He ordered Rob to stop talking with anyone about his daughter's hormone therapy except for the people the court authorized him to talk to. According to the judge's words, Rob was causing Mary "significant risk of harm" by "publicly rejecting his [sic] identity, perpetuating stories that reject his [sic] identity, and exposing him [sic] to degrading and violent commentary on social media."

Rob's next petition was thrown out by the court. He appealed to one of the highest courts in the land. They ruled that Rob's "refusal to accept [Mary's] chosen gender and address him [sic] by the name he [sic] has chosen is disrespectful of [Mary's] decisions and hurtful to him [sic]." In an infinitesimally small act of "mercy," while the court did allow Mary's treatment to continue, they lifted the gag order on Rob... but only when it came to his private communications. He was still barred from talking with the media.

Roughly three years had passed since the start of this journey, and Rob began to face up to reality. His daughter was largely lost. She was sixteen. She was "affirmed" by everyone in authority. The testosterone had even caused her to produce facial hair by that point. But by this time, the fight was bigger than one family. Rob knew he had to help other parents. He had to stop this coordinated program of child endangerment.

But the state was running out of patience. Rob never stopped speaking out. And the courts decided he needed to be made an example. Rob had never had trouble with the law. This was his only real interaction with the court system. Nonetheless, a judge ruled that enough was enough. In March of 2021, Rob was sent to prison.

When I read author Bruce Bowers's depiction of Rob and Mary's story in *City Journal*, it all sounded like a dystopian fiction. It isn't fiction, but it is dystopian—and the dystopia is in

America's backyard. This all happened in Vancouver, C<
mere thirty miles from the U.S. border.

Canada is perpetually one small step ahead of the Am
leftwing crusade. As a general rule, add a few years and w
happening in Canada will be happening in the United States.
legal imposition of the transgender regime is coming. But if)
talk to the advance guard of the transgender left, you'd nev
know. They don't mention gag orders on parents and priso.
terms for dissidents—at least not yet. Right now, everything is
sunshine and rainbows. It's easier to sell something if it's all benefits and no costs.

HOW TO MAKE DISAGREEMENT ILLEGAL

Nobody exemplifies this approach more than U.S. Representative
Mark Takano of California—lauded as the first openly gay "person
of color" to be elected to Congress. I sat across from Representative Takano in his district office in Riverside, California. He's a
member of the Equality Caucus and one of the original co-
sponsors of the so-called Equality Act, a transformative pro-transgender piece of legislation passed by the House of Representatives in February of 2021. I asked him about the bill, and he
responded with rather pleasant-sounding boilerplate.

"Well, very briefly," he said, "it's the most expansive and comprehensive civil rights legislation for LGBTQ Americans. It basically puts them on, I think, an equal footing in all fifty states and
the territories. Currently, that's not the case because it is still
possible, depending on where you live, to be legally discriminated against in a number of areas of the law. So public accommodations is one area, education is another, serving on juries,
credit applications."

Nothing sounds more positive than taking up the mantle of the Civil Rights Era. But this bill does more than copy the rhetoric of an early time; it co-opts the entire structure of civil rights law to promote the transgender agenda. "The most simplest way to talk about the Equality Act," Representative Takano continued, "is that it simply amends the 1964 Civil Rights Act to include sexual orientation and gender identity so as to include these two groups of people as protected classes."

In layman's terms, that means disagreeing with the trans agenda will be treated the same as being a racist. In the left's limited perspective, it sort of makes sense. After all, if being transgender is an identity, not a psychological disorder, and if that identity *must* be affirmed or else mass numbers of people will commit suicide, then it makes sense to ban "discrimination" against those people. Of course, the side pushing this agenda considers gender both an immutable identity like race when it comes to the law and a fluid social construct when it comes to individuals, which means it is nothing at all like race. Obviously, they want to have their cake and eat it too.

But that's not the only way the comparison breaks down. The 1964 Civil Rights Act was passed to ban discrimination that had no basis in reality. Skin color is a real thing, but it wasn't a vitally important difference when it came to the questions of the day like who can swim where or eat where or stay in what hotel. On the other hand, by attempting to erase the idea of biological sex, transgenderism erases a distinction that really does matter.

So while it is discriminatory to say that only whites can drink at this fountain and only black people can drink at that fountain, it is a perfectly reasonable request to say only women should use that locker room and only men should use this locker room. It

makes complete sense, based on biology and the physical differences between men and women, to say that women can have sports leagues only for biological women and men have sports leagues only for biological men. That's not discrimination. That's just common sense. And that's because sex is different than race.

By labeling real, useful, and necessary distinctions between men and women as "discrimination," the left is playing a shell game. They want you to believe that something totally normal is irredeemably evil.

The weaponization of the Civil Rights Act in this way will destroy all female spaces. It will send the message that those who believe giving cross-sex hormones to children is wrong are really no better than defenders of racial segregation. It will also force Americans to either act against their conscience or lose their jobs.

For example, if a pharmacist really did believe that giving cross-sex hormones to a minor wasn't "affirming" but was really medical experimentation more akin to child abuse, it would be discriminatory for that pharmacist to refuse to hand over the drugs. And we know that the activist left won't just ignore the few dissident pharmacists and find pharmacists who will provide them with the drugs they want. Just like what happened with Jack Philips who has been taken to court repeatedly because as a Christian he didn't want to bake a cake for a gay wedding or a gender transition celebration, and just like the Little Sisters of the Poor who didn't want to pay for insurance that included contraception (I mean, come on, they are *nuns*), the left purposefully targets for persecution those who disagree with them. That means no pharmacist would be safe. You either hand over the testosterone or find another job.

Pharmacists aren't the only ones with a target on their backs. Medical professionals of all sorts who disagree with medical transitioning would be targeted and possibly forced to go out of business. This is already happening in blue states. Catholic Hospitals in New Jersey and California have been sued because they refuse to perform hysterectomies on perfectly healthy women who say they are male. Another hospital in Washington State was sued because it didn't want to chop off the breasts of a physically healthy sixteen-year-old girl who wanted irreversible surgery to confirm her "identity" as a male.[2] The Equality Act would nationalize these lawsuits, and any doctor, nurse, or hospital system that fails to go along would be crushed by the law.

But when I asked Representative Takano if the Equality Act has any tradeoffs at all, systemically or individually, he completely shrugged off the concerns.

"Let's get into more specific policy issues. We talked about public accommodations, bathrooms," I said. "There are some women who say... 'I'd like some privacy in the bathroom. I'd prefer not to encounter naked penises, frankly.'" I decided to lay it on thick with Representative Takano to make him feel like I was on his side. "These people say even that the penis is a telltale sign if someone is a male. There are people who kind of really bought into the rumor that only men have penises," I told him. "How do we account for that?"

"Um, well, um," he started, auspiciously. "Well, what I would say is that most transgender people that I know... Um, I think a person who wants to use the women's bathroom who identifies as transgender really does think of themselves [sic] as a female. And, you know, part of how we can deal with the situation in the future is accommodating transgender people in public bathrooms."

They think of themselves as female. That's it. Interesting. That made me wonder if Representative Takano could answer the question that so far none of the pro-transgender side had been able to answer.

"So, what is the difference between sex and gender identity?" I asked him.

"Well, sex is clearly related to, you know, discrimination based on sex, I think most commonly people understand, is discrimination against someone who might be a woman. They've often been the disadvantaged gender group," he responded.

Huh. So sex can refer to women who are a gender group... It sounds an awful lot like all the buzzword terms were just being thrown around haplessly in the hope that I really didn't care about getting an answer to my question.

He continued, elaborating on the subject of gender identity. "There were cases that involved actually transgender people... who were dismissed because... they were not gender conforming... The Supreme Court found that unlawful. The reasoning kind of went that it related back to discrimination about, um, but it was about, it was clearly a gender identity sort of issue... They were discriminated against based on what they were perceived to be."

Ok, so gender identity is just a perception people have of themselves? Or a perception others have of them? It wasn't really clear what Representative Takano was trying to say. For a person who has made it his mission in Congress to ban sexual orientation and gender identity "discrimination," he seemed to have no clear idea what the words "sex" and "gender" actually mean.

I asked him what the words meant outside of what the Supreme Court has ruled, and he yet again took a left turn and started talking about something else, this time intersex people.

Again, that wasn't what I was asking, so I went back to the original question.

"For those of us who are not transgender," Representative Takano said, "it's, I think, pretty difficult to maybe wrap our heads around that, accept it. But I think I've come to understand it as a real identity [that] needs to be respected."

All right then. We can't understand what sex and gender are because we're not transgender. But even though we can't understand it and can't even describe it, transgenderism is a completely real identity that we need to respect. Yet again, I didn't get an answer to my question. But I did learn that Representative Takano was not the sharpest tool in the shed. I decided to move on, asking him how the Equality Act impacts public accommodations, like public restrooms.

"There are many different solutions to it," he said, having mastered the politician's art of vaguely responding by saying nothing at all. "There are ways to structure public accommodation of business going forward so that all can be respected."

Essentially, he's saying that this issue isn't a problem because he says it's not a problem. If people are uncomfortable, it's because we just haven't "structured" everything properly yet. I obviously wasn't going to accept this non-answer, but then, just like Professor Grzanka, Representative Takano decided he had had enough. He didn't like being asked basic questions about a law he voted to pass that will fundamentally transform America and destroy any recognition of sex differences in law. He was there to get good press, not to answer questions.

"We're going straight to the controversy over bathrooms," he said, "so you know what, I think this interview is over. Yeah, I think this is over." He quickly stood up, took off his mic, and walked out of the room.

Obviously Representative Takano didn't want to talk about the details of his legislation. He couldn't even acknowledge that a single person might be negatively affected by it. So, I decided I'd talk with someone on the other end of the political machine. Instead of a Congressman, I wanted to talk to someone in Washington who actually had power. I went to the head of an activist organization.

WHO IS UNCOMFORTABLE?

Rodrigo Heng-Lehtinen is the Executive Director of the National Center for Transgender Equality (NCTE), an activist organization founded in 2003 to advance the trans agenda in Washington. She presents herself as a male—and she came prepared with the similar cut and dried talking points that Representative Takano had.

First off, I asked Heng-Lehtinen what exactly her center does.

"We shape the laws and policies that govern our lives so that transgender people don't have to face... discrimination any longer." Just like Representative Takano, the push is to end "discrimination," automatically implying that distinguishing between biological males and biological females is wrong. That being said, apparently NCTE has been very successful in recent years.

"We were founded in 2003," she went on, "and notably when we first started, there were members of Congress who wouldn't even take a meeting with us... Now we've come so far that we actually got to interview President Biden about transgender rights and hear him speak really openly and affirming about transgender people. So we've been able to get a lot done in just that amount of time."

I asked her to expand a little bit on their activities, and she told me it's not just laws like the Equality Act that they focus on. They also try to influence the real power center of the national government—the federal bureaucracy. "You know when people think about government, they usually think of the White House, the Capitol, and the Supreme Court," Heng-Lehtinen said. "But there's actually all these agencies out there like HUD and the VA and HHS. Those might not be household names for everybody, but these government agencies are the ones who design all these rules that govern our lives. So, for example, with the Veterans Administration, they are the ones who determine whether transgender veterans can get health care. I mean, these are literally life or death issues."

Through this type of lobbying, NCTE can change all sorts of rules governing our society without our representatives ever having to vote. One of Heng-Lehtinen's proudest achievements on this front is in health care.

"A lot of people don't realize that it used to be really, really difficult to get health insurance if you happen to be trans just because you were trans," she said, which is obviously not true. Trans people have always been able to get health insurance, just like everyone else. They just couldn't have other people pay for medically unnecessary surgeries and therapies to affirm their self-perception—also just like everyone else.

Heng-Lehtinen continued. "There used to not be any kind of rules that barred an insurance company from turning you down because being trans is technically a medical condition. And insurance companies used to be able to reject you on the basis of a preexisting condition, which being trans counted as. So we had a huge victory when we supported the passage of the Affordable Care Act, otherwise known as Obamacare."

But the passage of Obamacare was apparently only the first step in the process. Once the legal structure was set up, NCTE worked to alter the rules from inside the federal bureaucracy to expand coverage of medical transitioning.

"But then furthermore, we've worked for years and years and years to strengthen it," she continued. "It is thanks to these policy wins that now an insurance company has to cover you... They have to cover transition related care. So it used to be that when you came out as trans, and if you needed hormones or surgery, they could deny you just because it was that you were trans, even if they covered the exact same procedures for people who weren't trans. So now they can't do that anymore."

Of course, a necessary mastectomy for a victim of breast cancer is not the same as an elective mastectomy for a fifteen-year-old girl who thinks she's a boy. In one, the medical intervention is done to improve the health of the patient. In the other, it's done to solidify a person's psychological disconnect from reality. That's not refusing to cover the exact same procedure just because someone is trans. That's refusing to cover the exact same procedure because in one case, the procedure was necessary and in the other, the procedure was not. But in Heng-Lehtinen's mind, to accept one and reject the other isn't a medical decision; it's an example of discrimination.

She made it clear that in her mind, medical transitioning isn't some personally desired cosmetic change as people like therapist Gert Comfrey described it. It is essential to health. "All doctors around the country have guidance—from the American Medical Association, American Academy of Endocrinology, and so forth, the American Academy of Pediatrics—every single leading medical institution has found that this is essential health care," she told me.

Of course, when Obamacare mandates coverage for gender transitioning, that means taxpayers like you and me are helping to foot the bill for it. The entire system is kept afloat by taxpayer-funded subsidies and kickbacks to insurance companies.

Just like Representative Takano, Heng-Lehtinen had no sense that there could be any conflict between trans "rights" and the rights of others.

"You do hear some women who say, 'I don't feel comfortable in a locker room with an individual who has a penis.' What do you say to those women?" I asked.

"Transgender women, just like anyone else, value safety and value privacy," she said. "Transgender women don't pose a threat, and in fact, transgender women are actually vulnerable to harassment themselves."

"So when someone says, 'Hey, why can't the trans woman just use the men's room?' What's the answer to that?" I asked.

"The reason that a transgender woman can't be told to just use the men's room instead is that it's not safe. Transgender people encounter so much harassment in daily life," she said.

"So we can't put trans women in the bathroom with men. It's not safe for them," I continued. "What about the woman who says, 'Well, don't put me in a bathroom with a man; it's not safe for me.'" I let my mask fall for a second, revealing that I believe in the truth that trans women are actually just men. Luckily, she didn't catch on.

"Well, you know, again, trans people value safety just like anyone else," she said, noting as well that states that have allowed transgender people into women's bathrooms have not seen an uptick in violence.

"Except for the Wi Spa incident, right?" I posited.

"Well, the Wi Spa case, I don't know too many specifics about," she said. "Except that I do know that it was debunked in some way."

It wasn't. I let it slide.

"I guess this goes to this tension though, right?" I said. "Because you have some women who say, 'I don't want to share a locker room with a male. It makes me feel uncomfortable.'... There's two competing rights claims, and really we would have to side with one or the other, but someone is out of luck, right? So you would say the women have to be out of luck in that scenario?"

"I would say that the more people get to know the transgender people in their lives, the more that discomfort will dissipate," she said.

"But now they're uncomfortable," I responded.

"And I would still say that that discomfort dissipates," she said.

Interesting. Then it begs the question: why are we forcing women, who have always had their own restrooms, to deal with this discomfort and not trans people?

"Would you accept that answer the other way," I went on, "if someone said, 'Well, you may be uncomfortable as a trans woman in the men's room, but that discomfort will dissipate."

"That discomfort won't dissipate because transgender people are subject to harassment when they are put into the wrong facilities," she said.

"Women are subject to harassment as well," I retorted.

But it didn't matter. The conversation was crystal clear. Yet again, we must be forced, by law, to enter their reality, but we are forbidden from asking them to live in actual reality. It doesn't matter what happens to women. It doesn't matter what happened to that six-year-old girl at Wi Spa. Transgender people rank higher on the victim totem pole. Their way wins, their harassment is deemed worse than harassment against others, and that's that.

Heng-Lehtinen effectively had the same answer on sports.

"Let's take a real-life case just to frame it, you know?" I said. "There was a case in Connecticut; there were two male track runners who..."

"They were trans girls," she interjected, catching on to me. I didn't care anymore. I was impatient with her obfuscation and assertions that all these legal changes have no negative impacts whatsoever.

"Right," I continued. "You look at those individuals, you look at their times against boys, and they were kind of middle of the pack. And then they race girls, and they're, you know, first and second place. Is that indicative of some kind of unfair advantage that those individuals might have against girls?"

"The Connecticut case is the exception," she said. "They got a lot of attention because those two trans girls performed well."

She then took it a step further. "We're now facing bills all around the country that are exclusively about banning transgender young people from either accessing best practice medical care or from being able to access school sports, and they're written such that they're only about trans youth," she said. "They're not even broadly about health care or sports; they're just about putting a target on trans kids backs... It is so heartbreaking to think about being an adult and thinking that you're going to put your time and energy into essentially bullying trans young people."

Bullying trans people? What about the trans movement bullying everyone else?! Heng-Lehtinen kept telling me over and over again that people's opinions on transgenders will change when they get to know transgender people. I wondered if her opinions would change if she met the six-year-old girl who had a penis shoved in her face at Wi Spa. I wondered if she would change her mind if she met Selina Soule who lost race after demoralizing race after being forced to compete against two biological males. I wondered if she would change her mind if she met Curt Schilling who lost his job and Scott Newgent who was blocked from social media and Don Sucher who was harassed in his own store—all because of the intolerance of the transgender movement.

Everything the pro-trans politicos were telling me sounded so positive and kind. It was about ending discrimination. It was about respect. It was about affirmation. Every policy and regulation and reform was a win-win. Who could be against it?

Yet all the while they were working to force, by legal mandate, the acceptance of biological men into women's spaces, to use our tax dollars to fund hormone therapy and sex change surgeries, and to mark any resistance to their agenda as no better than racism.

SLIDING DOWN THE SLIPPERY SLOPE

The truth is we're not that far from Canada. It starts with benign-sounding, anti-discrimination laws whose effects are far larger than proponents claim. But soon enough, kids who aren't affirmed will start being taken away from their parents. Scratch that. It's already happening.

A dad named Jeffery Younger in the Dallas-region fought for years to keep custody over his nine-year-old twins, one of whom was obviously indoctrinated into believing he is a transgender girl. Just like what happened in Canada with Rob Hoogland, Younger said that his child's school district was "right now actively teaching" his son to be a girl. "When I took James to school in boys' clothes, the teacher gave him a dress to wear," he said. "The school doesn't call James by his real name. They use a girl's name. They actively teach James that he is really a girl. They even make James use the girl's bathroom."

Remember, this is happening in Texas.

Eventually Younger lost custody to his ex-wife who later admitted that she may have "over-affirmed" her son's female identity. Thankfully, unlike the Hoogland case, the judge barred the mother from beginning hormone therapy on her son without his father's consent.[3]

Younger isn't the only one though. In Ohio, an unnamed couple was united in their opposition to their seventeen-year-old daughter undergoing hormone therapy in an attempt to become a man. But because the court found that this failure to affirm triggered suicidal feelings, a judge took custody away from the parents and gave it to her grandparents.[4]

Some states preemptively bar any attempts to prevent children from transitioning. Fourteen states and Washington, D.C. prohibit what is derogatorily called "conversion therapy."[5] That means that everyone—parents, counselors, teachers—are prevented from making an effort to help a person gain or regain a gender identity that conforms with their biological sex.

Of course, trans activists are engaged in a population-wide campaign of conversion therapy every single day. Transgenderism is in schoolbooks, on ads, placed into TV shows and movies, and pushed by teachers and therapists and doctors. Every attempt is made to get young people to explore their gender identity and embrace gender fluidity. Every movement that contradicts a person's biological sex—from acting out the most basic stereotypes to actual clinical feelings of gender dysphoria—are affirmed and embraced and cultivated ad nauseum with or without parents' consent.

Yet even the smallest attempt to counteract this and assert that biological sex is not some random, useless, outdated concept is treated not merely as backward, but as illegal. And you can guarantee that trans activists like Representative Takano and Heng-Lehtinen would take the first opportunity to ban so-called conversion therapy nationwide if they had the chance.

All of this is more than a case of attempting to end discrimination. It's the enforcement of the idea that gender is an immutable identity and that any disagreement with that cannot be accepted.

Those in authority will go to absolutely insidious lengths in order to strip parents of their rights and spread the errors of transgender ideology. One father in Washington State saw this clearer than most anyone.

In order to hide his identity, he is only known as Ahmed. He's a Pakistani immigrant to America, and the state came for his son—just like how the Canadian state took away Rob Hoogland's daughter.[6]

In the fall of 2020, Ahmed admitted his sixteen-year-old autistic son to the hospital in Seattle after his son threatened to commit suicide. Within days, the hospital emailed him saying he should take his "daughter" to the gender clinic, using a new name for his son as well.

"They were trying to create a customer for their gender clinic," Ahmed told journalist Abigail Shrier, "and they seemed to absolutely want to push us in that direction." You won't be surprised to learn that the counselors and therapists assured Ahmed that the only way to stop his son from being suicidal was to affirm his new gender identity.

The laws in Washington were already working against Ahmed. Minors can get gender-affirming care without their parents' permission at only thirteen in the state. Ahmed knew that he had no power in this situation, so he talked to some trusted friends—a lawyer and a psychiatrist. The psychiatrist told him the dark truth behind the transgender takeover: "You have to be very, very careful," he said. "If you even come across as just even a little bit anti-trans or anything, they're going to call the Child Protective Services on you and take custody of your kid." The lawyer agreed that the only way to get his son back was to agree with the hospital. It was the only way to get his son home.

Ahmed did as he was told... to an extent. Once he successfully got his son home after assuring the hospital that he would take him to the gender clinic, Ahmed confronted a decision. He could either face the potential loss of custody of his son if the state found out he didn't follow through. Or he could flee somewhere where the laws weren't yet as radical.

Ahmed chose to save his son. He quit his job, moved his entire family out of Washington, and went somewhere where parental rights are still respected. Take that in for a moment. An American man was forced to pack up his family and flee oppression, not from some foreign land, but within America itself.

Ahmed's story ended well. But people who want to save their children are running out of places to hide. Similar laws as Washington govern California and Oregon, meaning that parents are nearly powerless to prevent their kids from transition anywhere along America's western seaboard.[7] It's only a matter of time before these types of laws creep across the entirety of blue America. If the trans ideologues had their way, Washington's laws would be federal policy.

What the left is erecting has come to be referred to as, using the words of writer James Poulos, the pink police state. It is totalitarian in nature but warmly wrapped in an ever more multi-colored rainbow flag. The pink police state won't be enforced by grim, army-green-wearing shock troops. Its soldiers are political activists, teachers, and members of Congress. It enforces order with lawyers and through your fellow citizens, ever at the ready to film your words and movements in an attempt to catch any whiff of anti-trans sentiment so that they can post the video on social media and turn the outrage mob against you. It keeps people in line, not at gunpoint, but through the resident Nurse Ratched who informs you with a cold, polite smile that your daughter is now your son, and we would hate to see what happens if you don't take him to the gender clinic immediately.

They don't claim to want a new world order. Their rhetoric is understated. They just want to end discrimination. They just want acceptance. They just want respect. But their actions and laws and rules reveal grander ambitions.

Their regime of anti-discrimination discriminates against women and attempts to end any distinction between male and female. Their desire for acceptance demands affirmation and bans dissent. They require your respect, but they refuse to re-spect the dictates of biology, the innocence of children, or the most elemental and sacred bond in all of human existence—the bond between parent and child. Their words are easy, but their burden is large.

But we aren't without hope. There are men of bravery like Ahmed and Rob Hoogland who would rather sacrifice themselves than sacrifice their children. There are strong women like Selina Soule and Scott Newgent who speak the truth, no matter the conse-quences. The transgender blitzkrieg has taken over our nation and our culture with astonishing speed. But the resistance is growing.

THE REBELLION

SO... WHAT IS A WOMAN?

Can you answer that question? You should be able to. Everyone should. But against all odds, it has become an extremely thorny question.

It's not because the answer is complicated. It's not because we've learned something new about womankind that no previous generation had known before. The only reason people have trouble answering this question is because an angry, vindictive, nonsensical ideology has taken over. As a result, people who know the answer are afraid to say it. People who should know the answer have gone so far down the gender theory rabbit hole that they have ceased to be able to speak with any sense.

I had spoken with a therapist, psychiatrist, pediatrician, clinical psychologist, surgeon, activist, athlete, congressman, professor, store owner, and a transgender person. We talked about gender theory, John Money, the Reimer twins, rapid onset gender dysphoria, hormone therapy, sex change surgery, high school sports, the Equality Act, and so much more. I learned more about vaginoplasties than I ever wanted to. I had to say the word "penis" more than I ever thought I would. But there was still that one, single question I needed to ask unambiguously and directly.

CAN YOU JUST ANSWER THE QUESTION?

"What is a woman?" I asked my trans-affirming therapist friend Gert Comfrey.

"Yeah, great question. I'm not a woman, so I can't really answer that," she said.

Well, ok then.

She continued, "No two women will answer that question the same way."

I probably shouldn't have expected anything more from a person who explicitly condemned the idea of absolute truth. I gave it one more shot anyway.

"Would you say there is no definition of 'woman'? Really, it's not a word that means any particular thing?" I went on.

"I think it's relative," she said.

That was strike one for me. I turned to my next interlocutor, the pediatrician who prescribes hormones to children, Dr. Michelle Forcier, hoping for some better luck.

"What determines someone's sex?" I asked.

"It's a constellation," she said.

Sigh. Ok, let's try this again.

"What is a woman?" she asked rhetorically. "A woman is someone who claims that as their identity."

"You're using the word 'woman' in the definition for a woman," I answered. "It's like if I asked you what's a tree and you said, 'A tree is a thing that's a tree.' You see, you haven't told me anything about trees."

"I'm not here to talk about trees," she responded.

How long shall I be with you? How long shall I suffer with you? Jesus's words in the Gospel of Mark came to mind.

"It's an analogy," I said, dryly.

"Yeah, but your analogies don't work," she told me.

"I'll try one more time," I went on. "What if you were to define the word 'woman'? How would you define it?"

"I would ask the patient, what is their definition of 'woman,'" she said.

No wonder Dr. Forcier's patients are so confused about sex and gender. Their own doctor has no idea what she's talking about.

Transgender activist Rodrigo Heng-Lehtinen was no different. "I hesitate to go too far down the terminology stuff because it can kind of distract us," she told me. Well, the only reason it's a distraction is because nobody on a certain side of the debate seems to have any idea what they're talking about, I thought to myself. She continued. "I mean, really, a woman is someone who says that she is a woman and transitions to be a woman."

"A woman is someone who says she's a woman," I began, "but what is she saying she is?"

"Well, you know, if someone says to me, 'I'm a woman transgender'... then I respect that, and I know that she deserves to be treated with respect and with dignity and care no matter what, just like anybody does."

Of course, I don't disagree with that. But that is hardly the question I asked. Heng-Lehtinen may have mastered the art of the political pivot, but I had a single-track mind at this point. I was going to get a real answer somewhere. So, I kept going down the list.

I wasn't exactly looking forward to asking Women, Gender, and Sexuality program professor Patrick Grzanka the question. He had already accused me of denying people's existence at that point. But there was no turning back now. "So, what is a woman?" I asked directly, just like with everyone else.

"Why do you ask that question?" he shot back, suspiciously.

"I'd just really like to know."

"What do you think the answer to that question is?"

We went back and forth as he grilled me, asking me why I was so interested in this particular subject. He was acting extremely curt and looked at me with profound mistrust, as if I were asking him where I could find illicit drugs or if he had just a few minutes to sign a petition to save the environment.

"I wanted to answer questions about women's studies," I said. "And so the first answer you should be able to provide is what exactly is a woman?"

"Well, for me, it's actually a really simple answer," he said, "and that's: a person who identifies as a woman."

Apparently circular reasoning is a feature of transgender ideologues. They either claim they can't answer the question because they lack a certain identity, or they define the term using the term itself, making the definition completely meaningless.

"You are seeking what we would call in my field of work an essentialist definition of gender," he said, twisting my words.

Actually, I'd simply asked what the definition of a word was. As far as I'm concerned, gender theory didn't need to have anything to do with it. He went on, "I think it sounds like you would like me to give you a set of ideological or cultural characteristics that are associated with one gender or the other."

"I'm not seeking any *type* of definition," I answered. "I'm just seeking *a* definition."

"And I gave you one," he said, quickly.

Alright then. I felt like I had just about struck out. But I soldiered on.

I next turned to sex change surgeon Dr. Marci Bowers. Say what you will, he was one of the saner members of the transgender ideologue crowd. After all, he was one of the only people I spoke with on that side who admitted that gender transitioning actually does come with some risks.

I let the question drop, anxious to finally hear an intelligible answer. He began, "Well, again, this is... it's a, it's a... so a woman or a man. What is a man? A man is someone who, uh, again, it's not so much even, uh, you can have a male gender identity without necessarily being a man."

Yeah, that didn't clear up anything. Everyone caught up in gender theory seemed incapable of answering a really basic question. Perhaps surprisingly, the most honest answer I got from this crowd came when I was in Representative Mark Takano's office—only it wasn't Representative Takano that said it.

The California congressman had already decided to storm out of the interview before I was even able to ask my ultimate question. As he got up and hurriedly rushed out of the room, I tried to get it out. "We want to know, what is a woman?"

"Please turn off the cameras," he said.

"We came all this way," I said as the congressman began to leave the room with his staffer a few steps behind him. "We just want to know what is a woman?"

"And you're not going to find out," the staffer said before closing the door.

The staffer was wrong in principle, but she was right in one respect: I wasn't going to find out by talking to trans ideologues. Everything that came from them were non-answers, circular reasoning, and long-winded monologues that went nowhere.

Their answers, however, did underscore an important lesson. These people really don't care about the truth. This isn't ultimately about doing what's right—since you can't have right or wrong without some semblance of the truth. This is all about advancing an agenda. If that weren't the case, they wouldn't be continually hiding something and using euphemisms.

There is a good rule to follow in life: You can tell if someone is telling the truth if they aren't trying to hide anything. They let their yes be yes and their no be no. Clarity is a mark not only of sincerity, but honesty.

That's how I knew that the others I had interviewed weren't lying to me. They didn't mince words. They didn't try to change the subject or pivot to answer a question I didn't ask. They told me the truth, not necessarily because they were better educated or more powerful or had better credentials than any of the other people I had spoken with. No, it's because they chose to never forget—and never be forced to forget—what everyone before us in history already knew.

"What is a woman, exactly?" I asked Selina Soule, the track runner who was repeatedly beat out by biological males during high school races.

"A woman is somebody who has the reproductive organs to give birth to a child. That is what a woman is. There is nothing that can be done to change that," she said.

"It's as simple as that?" I asked.

"It's as simple as that."

Wow, I hadn't gotten an answer like that before. Perhaps I could go two for two. So, I put the same question, word for word, to Scott Newgent, the woman who related to me the terrifying story of her attempt to medically transition from a female to a male.

"What is a woman?" I asked.

"It's genetics," she said. "It's chromosomes. It's factual. It's biology." She didn't let it rest there. "I can create whatever kind of version I want of female, but once we lose reality, what do we lose? Our minds. We're losing our minds."

Dr. Grossman, a child psychiatrist, remained, as always, a bastion of sanity in the chaos of gender theory. "What is a woman?" I asked her, confident she would tell it to me straight.

"It's not a complicated question. It's been turned into a very complicated question, but it's not," she said. "A woman is a female. A female is a biological being with two X chromosomes who reproduces by means of a gamete called an egg, who, in most cases, can conceive a child naturally and carry it in her womb. That's a woman. Simple enough, right? We don't have to have pages and pages of explanation about what is a woman and what is a man. It's not a thorny, complicated subject."

Science—real science, not the quack pseudoscience peddled by transgender activists—is exceedingly clear.

But I have to admit, my favorite answer came from Mr. Don Sucher, the owner of the *Star Wars* memorabilia store in Washington State who was accosted by a transgender person for the crime of acknowledging that women can't have penises.

"I don't have to define a woman," he said. "They're either male or female. Yeah, that's it. That's how you were born."

"That's all?" I asked.

"That's all there is to it. Plain. Bottom line."

I didn't need to ask Sucher any more questions than that, but I couldn't help myself. Something told me he'd give a beautifully colorful answer if I continued, and I was rewarded in my hope.

"How do you know that you are a man?" I asked him.

"How do I know that I'm a man? I guess because I got a dick."

The everyman got it right when so many of the powerful and educated made no sense at all.

In a way, that gave me hope. The forces of politics, law, business, entertainment, education, academia, and medicine are arrayed against us. But all their so-called wisdom crumbles in the face of even just one man willing to tell the truth—no exceptions, no sugarcoating, no apologies.

To put it in terms Sucher might prefer, we're like the Rebellion in *Star Wars*. We have less funding and less power. We are certainly less ruthless. But we have what is true and good on our side. We have basic human decency and common sense. Together, that not only counts for something, it can count for everything.

THE TRUTH IS MARCHING ON

To overcome the quickly forming transgender tyranny, we need to know how to fight back. Little did I realize in the moment, but in their brave examples, people like Selina Soule, Scott Newgent, Jordan Peterson, Dr. Miriam Grossman, and Don Sucher were showing me how to do just that the entire time.

The most important lesson I learned from their examples was to always tell the truth, no matter how much the other side bullies you. Compromising on the fundamental facts of biology and language doesn't help find common ground with our opponents. It is seen as a sign of weakness. If they're willing to give hormones with long-term side effects to thirteen-year-olds while stripping parents of custody of their own child when they don't agree with what's going on, they certainly aren't going to be satisfied if you offer a simple pinch of incense to Caesar by using preferred pronouns without agreeing with them on everything else. There is no cafeteria transgender ideology here. As J.K. Rowling taught us, you either buy into all of it or you are condemned. They do not accept half measures.

That being said, our heroes knew that the moment they spoke the truth, they put a giant target on their backs. But they didn't hide from the fight, hoping that by not choosing sides they could get away unscathed. Some, like Sucher, chose to pick a fight. When he put that sign up in his shop, he didn't do it to make people angry. But he had to have known that transgender activists would come for him. He did it anyway.

Others like Selina Soule never wanted to get involved in this issue in the first place, but the fight came to her. She could have been like almost all the other girls running track in Connecticut. She could have privately complained but said nothing about it. Instead, she spoke up about the injustice of biological males playing in women's sports, knowing that by doing so, she would face vicious attacks.

Remember, Soule and Sucher aren't politicians or public personalities. They are just a student and a store clerk, respectively. Fighting the forces of the radical transgender left isn't exactly part of their job description. But they recognized there is more to life than comfort and anonymity. They knew that gaining the respect—or at least avoiding attacks—from the powers-that-be meant nothing if they aren't even allowed to state a basic, observable, clear-as-day biological truth. There should be millions of people like them.

To anyone who is still afraid, it's helpful to know that getting attacked by the trans activist mob didn't destroy Soule and Sucher's lives. While the attacks against them certainly hurt, both of them actually came out stronger in the end.

For Soule, she could have focused on the angry social media posts and the death threats, but she ignored them. "These people are just hiding behind a phone screen," she said, "and if they saw me walking down the street, they would never say these things to me." She told me that despite all the negative attacks, about 85 percent of the responses she got were positive. "And on the plus side, I received some marriage proposals as well," she told me.

Sucher reported an identical experience. "You know, I've gotten thousands of phone calls from all over the world," he said. "I would like to thank everybody, my God. But I'm never going to be able to do that in my life. But they are like 95 percent behind me."

All that makes me think of another lesson: We have many more allies than we realize. Who would have thought that one of the greatest defenders of sanity and children would be a biologically female, transgender person who went through the entire sex change process? Scott Newgent remains a lesbian and on so many counts would be considered a liberal, but on this issue, we were in 100 percent agreement.

Newgent actually learned for herself how many allies we can have when she was looking for help to defend children from the gender-transitioning medical mill.

"I found support in a lot of different places that I don't think I ever would have looked at without this experience," she told me. "I've had some famous evangelicals reach out to me that I at first wouldn't talk to. But then I decided, hey, you know what? It's important to stop medical transitioning. I talk to them and realize that, you know, we've come to some kind of conclusion that this is your boundary, this is my boundary. [Now] let's save kids."

Newgent told me she was surprised to be greeted with such open arms by the evangelical community, "They were just awesome people... and I fell in love with them. I absolutely did," she said. "When I have evangelicals reach out to me... I grab their hands and I go, you know what, it's OK that we don't agree on homosexuality, OK? Keep to yourself on that; I'll keep to myself on this. Let's save kids."

This cause is truly bigger than right or left, secular or religious, Democrat or Republican. Newgent's experience proves it. We may not control the culture makers or media outlets. But we are the majority. And we should never forget that.

People like Newgent and Dr. Grossman also reveal the power of knowledge. Every time I presented a claim from the pro-transitioning side, they came back with exact and specific criticism. They knew the studies. They knew the errors. They knew what had been debunked and what was true. They had studies of their own. They didn't just respond with stories or theories. They responded with documented facts.

We can know that the transgender ideologues are wrong seven days a week, but if we don't know *why,* our convictions are impotent. The other side has built an entire structure on a foundation of lies. Transgender people are *not* more likely to commit suicide if they aren't affirmed. Hormone blockers *do* have long-term side effects and are not reversible. There are *real and meaningful physical differences* between men and women.

It's easy to look at counselors and medical professionals who poison the minds of children and push those children to commit to medical transitioning and to react to it all with disgust. But we can't count on disgust alone to win the day. We have to be able to describe clearly why it's so wrong.

Of course, being able to say what's wrong doesn't matter if nobody hears us. The transgender activist Rodrigo Heng-Lehtinen may be promoting a tyrannical agenda that harms children, but we have to at least admit that she and organizations like the National Council for Transgender Equality are immensely effective.

That's because they don't let up. They push for legislators to pass laws, *and* they continually influence the bureaucracy to make the regulations around those laws more in line with their ideology.

Trans culture warriors are everywhere, not just in Washington. They are on social media and on TV and in movies and sports. We may not be able to match their reach in the short term. But the world has to know that there are millions of people—if not more—out there who reject gender theory wholesale. Part of that means speaking up personally and supporting people like Newgent and Dr. Grossman who already have established platforms. But we should also stop supporting, even tacitly, those who are trying to corrupt our children and infect the next generation with this ideology.

That means if your kid's school is pushing the trans agenda, you need to speak out. If they won't listen, you have an obligation to take your kid out of that school. Limit social media. Turn off television shows and movies that peddle this propaganda. Make your elected leaders know voting for this ideology is unacceptable. We can never support this pernicious ideology, not with our ballots, not with our time, not with what we watch, and not with where we send our kids to school. Our opponents have been taking a firm stand for years to push this on us in every front. It's our responsibility to dig our feet in and say "no."

It's not just our public actions that matter. It's our private character too. Over my journey, I saw that outside of the infinitesimally small portion of the population that experienced gender identity disorder from an extremely young age, those who experienced gender confusion later in life often had some sort of brokenness in their homes. It wasn't a 100 percent rule. But it happened too often to ignore. Most of the time, it was that the parents were divorced. Sometimes there was abuse. Other times, the parents were so leftwing that their entire family was completely unmoored from any tradition, religion, or existential stability.

Newgent noticed it as well. "Why is it that we have the kids that are being abused that are in homes and stuff—those kids have a higher chance of being transgender. So we have all the kids that we're supposed to protect in society and [we're] telling them that all you have to do is medically transition and you fit in."

Acknowledging these facts imposes a two-fold duty on parents and adults in the world. First, you can't imagine your private actions won't have consequences for those around you. Problems in your marriage pass down to your children. Instability in the home leads to instability in the lives of children. Divorce especially creates a massive disruption and vulnerability in the lives of kids where transgender ideology can enter in with its false promises and comforting lies.

Second, we have a responsibility to help young people in the world recognize that there are consequences to their actions, that transitioning is not cost-free and that this ideology is harmful. To use Newgent's words, we must help them "see around corners."

"That's what parents are, right?" she told me. "We see around corners. We love the hell out of our children, and we help them see around corners until they can see around corners themselves."

Newgent told me a story to show me that we have to be understanding with kids and help them because it's not their fault that they don't know any better. "When my kids were about seven years old—I had twins—we were at a pool party," she told me. "There was a baby that was just doing the toddler walk... And my twins were just obsessed with trying to watch this toddler. 'Don't get in the water. No, no, no. Don't get in the water.' And I watched them, and I was so proud of them."

Then, Newgent told me how this all relates to gender theory. "Later that night I was like, 'Why were you guys so concerned about [the toddler] Rhett around the pool?" They told her that they didn't want the baby to drown. And why would he drown? Well, because he's a baby.

"I tell parents the same thing [when it comes to gender theory]," Newgent told me. "Rhett didn't know that he was going to drown because he was a baby. It's not because he was dumb or there's something wrong with him." Helping adolescents see around corners works the same way. They have no idea that adults at school or in medical facilities or online will prey on them. They have little to no preparation to be able to resist when an authority figure tells them they are the wrong gender or that medical transitioning is completely safe. They are defenseless, and it's our job to defend them. And that job is more important now than ever before.

Certainly, we have to pay particular attention to what is being taught in our children's schools when it comes to trans ideology. But it's so much more than that. As we saw with rapid onset gender dysphoria, the lies of gender theory can spread like a wildfire through social groups and especially on social media. Educational materials, books, activist teachers, entertainment programming, and the like prime the pump, so to speak, by introducing children to these concepts at a very young age. Almost without knowing it, they learn the transgender vocabulary, and in ways large and small, it begins to shape how they see the entire world. That means when they face any sort of confusion brought on by puberty, a breakup, their parents' divorce, or any sort of trauma, many children are already predisposed to interpret their difficulty using the language and lens of gender, and they have an army of adults in the real world and online willing to confirm and cultivate that perception.

You can't be a helicopter parent, but you can be vigilant. And you can preemptively counter the assault on your child's psyche by teaching them the truth early and often.

I know that some people out there might feel it's already too late, if not in the culture at large, then maybe in their own families. For every Ahmed who successfully saved his son from the clutches of the trans medical machine, there are parents like those in Ohio, Texas, and Canada who have watched as their children have been taken away from them, maybe not by force of law, but certainly through the overwhelming allure and power of the gender propaganda.

But even if your son or daughter has already been infected with this brain parasite, you can't give up hope. One mother in California showed how it's done.[1]

Writing with the pen name Charlie Jacobs, this mother wrote of how her ultra-feminine daughter suddenly took a turn at only twelve-years-old. She was immersed in anime and cosplaying, which Jacobs didn't realize at the time had all sorts of gender-bending and sexual themes. Her daughter's public school then indoctrinated her with leftwing sexual propaganda, and all of a sudden the young girl started talking with her friends about their sexual identities. None of them wanted to be labeled as "basic" or a straight girl.

Jacobs then began to see what we now recognize as the telltale signs of rapid onset gender dysphoria. Her daughter broke off old relationships, spent more and more time online, and even set up fake social media accounts so that her mother wouldn't find out whom she was really talking to. After meeting an older, more mature non-binary teen, Jacobs's daughter cut her hair, stopped shaving her legs, and even started wearing boys' underwear. It was only a short matter of time before she announced she was transgender and, of course, started to threaten to commit suicide.

Eventually, Jacobs got the passwords to her daughter's real social media accounts. Almost everyone her daughter was talking to was a stranger. People were sending each other sexual videos. Children were discussing erotica, including subjects like incest and pedophilia. They talked about the sex change surgeries they wanted to get. Older girls were coaching younger girls how to sell nude photos of themselves for money. The dark underbelly of the sexual revolution had infested this young girl's life through her phone.

This is where so many parents don't know what to do. It seems like every force in the world is against you. How can you save your own child if she's already this far gone—being affirmed by those around her while she distances herself from you and calls you a transphobe for refusing to use her new name, address her by her preferred pronouns, or take her to get hormone injections?

That's when Charlie Jacobs went nuclear.

She deleted every social media app from her daughter's phone. She blocked her daughter's access to the internet. She deleted every contact she had and changed her daughter's phone number. She observed attentively her school's online classes. She threw away every piece of anime in the house, locked up remotes to the TV, and banned any friend who wasn't completely upstanding from seeing her.

Jacobs reported the reaction: "She hated me like an addict hates the person preventing her drug fix. I held my ground, despite the constant verbal abuse."

But Jacobs didn't end there. She read up on trans ideology, talked to real experts, and committed to only using her daughter's real name and actual pronouns. She wouldn't for a second buy into the lie or confirm her daughter's delusion. She then forced her daughter to listen to podcasts that rejected transgender ideology on their drive to school, including podcasts with stories like Scott Newgent's about trans people who regretted their transition. She filled her house with books and literature that revealed the truth about transgenderism.

At the same time, Jacobs endured brutal pain. For a year and a half, her daughter responded angrily and viciously. But with almost Christ-like meekness, Jacobs endured it. She waited for every single moment of vulnerability and openness to show her daughter unconditional love, and she bit her tongue every single time an angry word was about to come out.

After all this effort and pain, Jacobs slowly brought her daughter back from the brink. It was like the demon that possessed her slowly left. Her battle isn't over. But her daughter is coming back. Her daughter is still her *daughter*. And in a world like the one we live in today, that means something.

TAKING MY OWN STAND

Jacobs and Ahmed, Dr. Grossman and Newgent, Sucher and Soule—these are the names of heroes. In their own ways, they are sacrificing themselves ultimately to help others. They knew the truth. Heck, the ones I spoke with personally actually knew quite easily how to answer the question "What is a woman?" But they also defended the truth.

Honestly, their examples made me a little reflective. Over these past few months, I had uncovered the dark truth behind gender theory. I learned where it came from. I saw what it was doing. I became terrified by its ambitions. After all of this, could I really ignore my own advice? I know the truth. But I needed to act on it.

My opportunity came when Loudon County, Virginia, announced new policies to not only allow biological boys to use the women's restroom, but also to force teachers to affirm students' gender identity by using preferred pronouns. Just like the worst policies I had seen enacted in places across the country, the school board banned teachers from telling parents if their child had assumed a new gender identity.[2]

This was all happening as I was interviewing people for this book, and I decided that I couldn't be silent. I had to take a stand. I flew in and helped gather a crowd of protestors outside of a meeting of the Loudoun County School Board.[3]

The board did everything they could to stop me from coming. When they found out that I wanted to speak to them at their session directly, they instituted new rules saying that only local residents could speak. The rules went into effect the day I arrived. So, I decided to rent an apartment in Loudoun County. And really, the school board couldn't complain. "I've lived in Tennessee," I told a local reporter. "I've felt sort of like a Virginian trapped in a Tennessean's body. I identify as sort of state-fluid I guess."

I was able to get in the door, but they cut off the livestream feed for outside speakers, and they made me wear a mask even though I was standing dozens of feet away from them at a microphone. It didn't matter though. My message was going to be heard one way or another. I gathered my notes, approached the lectern, and spoke from the heart:

"I want to thank you all for allowing me to speak to you tonight, but you tried not to allow it. Yet here I am now. You only give us sixty seconds, so let me get to the point. You are all child abusers. You prey upon impressionable children and indoctrinate them into your insane ideological cult—a cult which holds many fanatical views, but none so deranged as the idea that boys are girls and girls are boys. By imposing this vile nonsense on students to the point even of forcing young girls to share locker rooms with boys, you deprive these kids of safety and privacy, and something more fundamental, which is truth. If education is not grounded in truth, then it is worthless. Worse, it is poison. You are poison. You are predators. I can see why you try to stop us from speaking. You know that your ideas are indefensible. You silence the opposing side because you have no argument. You can only hide under your beds like pathetic little gutless cowards hoping we shut up and go away. But we won't. I promise you that. Thank you for your time, and I'll talk to you again very, very soon."

You can bet on that. You can bet people like them will be hearing more from all of us. That's because we now know the truth, and inaction is no longer an option.

Gender theory is evil, and what it is doing to children is evil. It is a monumental contradiction devoid of logic and absent of love. If you truly cared about people, you wouldn't indulge in their self-destructive fantasies. You wouldn't promote the mutilation of children. You would help people embrace the truth, and you would protect our kids, no matter the consequences.

The smug ideologues forcing this insanity down the throat of our society may think they have the upper hand. They attempt to rule by propaganda, fear, and force. But they have awoken a sleeping giant. We are not afraid of them—and we will take down their degenerate reign of terror.

AFRICA

ABOUT FIVE HOURS into the drive we hit a rough, bumpy dirt road. We still had an hour to go, but honestly the dirt road wasn't that much worse than the paved highway we just left. My mission had brought me here, so I wasn't complaining. In fact, I was excited. I was far away from everything: Far from civilization, far from air conditioning, far from running water, cell service, and people who take pictures of themselves. Most importantly for my purposes, we were far, far away from the west. I was on my way to visit the Massai people who live almost exactly the same way as they have for generations. Straddling the border of Kenya and Tanzania in Africa, these people live in mud huts with thatched roofs, tend cattle, hunt, and teach their children to do the same.

As I had learned over the past year, gender ideology is a creation of the modern west. It masks itself as something rooted in science and descriptive of the natural condition of humankind. But in reality, it is a fantasy world with a made up language and made up rules. As I soon saw, nothing reveals just how unnatural and counterintuitive gender theory is than speaking with people who haven't been instructed in its language through the forces of mass media, popular culture, and the long reach of a homogenous education system.

That's why I was in Africa. I wanted to talk to people who had never heard of sex change surgeries and hormone suppressors. I wanted to talk to people who hadn't yet been instructed by "experts" on the difference between "gender" and "sex." I wanted to talk to people who would never think to put their preferred pronouns in their email signatures because, well, they don't have email. Maybe then I could see firsthand just how innate—or not—all of this gender theory stuff really is. Maybe here I could see if the gender theorists were as crazy as they seemed, or if I was really the crazy one.

My translator (who undoubtedly called himself "Paul" so that westerners like me would know how to pronounce his name) guided me into the village. I was welcomed with great openness and kindness. Paul led me over to an elder in the village surrounded by a group of men and, after some pleasantries, I asked him what, in his mind, is a man. He didn't seem terribly perplexed by the question. The translator responded, "To be called a man in our community, you need to have a knife, secondly, a spear, a stick, and to make your own village by fencing. And you go to marry and, having a family, that you can sustain them... to have cows so that the children have enough milk.... You need to sustain your community."

To them, manhood seemed to be determined by certain roles someone plays, which is something a lot of gender theorists would agree with. So I asked the men if a man could just decide to do the roles of a woman. "In Massai community he cannot," my translator relayed.

"Can a man become a woman?"

"No."

"What about a transgender?" My interpreter looked at me not understanding. "Transgender," I repeated.

232

He tried explaining the concept to the Massai man I was speaking with.

"No... if you want to become a lady but you're a man, you have something wrong in your mind, something wrong in your family, something wrong in you."

"What about if someone is non-binary?" I asked.

Paul looked at me, his brow furrowed in confusion.

"You know, non-binary..." I went on. It struck me that he had no idea what "non-binary" meant. And why would he?

"You're not a woman, you're not a man?" he asked.

"Yes, someone is neither. They're something else," I said. I realized that when you have to put these ridiculous concepts into basic words, it doesn't actually make any sense. Gender theorists have created an entire dictionary to describe things that nobody else in history ever thought needed to be described.

"He's saying we have never seen things like this," Paul said.

Maybe I moved into the gender theory lingo a little too quickly. I decided to go back to the fundamentals. "How do you know if you're a man?" I asked.

I realized that when I had first asked them what a man is, they responded telling me a man's role in society not because gender is only a matter of fulfilling certain duties. Rather, they presumed I was talking about social roles because the biological reality of being male or female was so obvious to them. The idea that people could believe they are a different sex than how they were born simply didn't make sense. Once they knew I was asking about something more fundamental, they didn't hesitate to answer.

"When a newborn is born, we identify quickly because a man, he has a penis. A woman, she has a vagina."

233

I spoke to a group of women as well to see if they had the same understanding of gender as the men. The group of women I spoke with were all beautifully adorned in jewelry, so I used that as a jumping off point to take things from a different angle.

"In my country," I told them, "there are men who will put on women's clothes and say they are a woman. What do you think of that?"

Paul explained the situation to them, and the women spoke back to him for a moment. "They say they have never seen or heard something like that."

"What is a woman if you had to give it a definition," I continued.

Paul began translating for the woman, "She's saying for a man, no breasts. Secondly, their private part is different because a woman has a vagina and a man has a penis.... And also a woman delivers, a man cannot."

"This is going to seem maybe a little shocking," I responded, "but in my culture there are people, women, who chop off those parts of their body, chop off their breasts. Does that make them a man?"

She wasn't at all convinced. "Let's come now to the time of having sex," she said. "Once you have sex with a man, you expect him to have a penis."

Well, it's pretty hard to argue with that. Talking with the Massai, it was clear that the sexual binary between men and women made sense to them. Non-binary and transgenderism didn't make any sense. And the truth is, it was terribly difficult to explain transgenderism to people who take the world at face value—who see very clearly that there are two sexes and that is that. As soon as you try to tell people like this that they shouldn't believe their lying eyes, you can feel just how empty and vacuous the entire edifice of gender ideology actually is.

For all the talk about "gender" being a social construct, the truth is that gender theory is itself one of the most complicated and monumental social constructs there is. A whole Jenga tower of critical philosophy, language, culture, and medical advances that allow once unimaginable manipulations of the human body all must first be built up to sustain the fiction of gender identity. The moment you step out of the mind prison socially constructed all around us, you see just how much gender ideology is a product of western privilege, luxury, and decadence. Without a society that can support the likes of professional sexologists, phalloplasty surgeons, gender studies professors, and queer affirming counselors, gender ideology would have never been invented.

Even asking about it made the Massai people look at me like I was an alien or a freak.

"What if it's a woman with a penis?" I asked the group of men.

"What?!" Paul quickly replied, leaning in and looking at me to make sure I hadn't misspoken.

"Where I come from in America, lots of people say I have a penis but I am a woman."

The entire group immediately started laughing. They weren't mocking me. They just couldn't help but laugh at the absurdity of what I just said.

"They say they are just laughing because they have never seen or heard something like that," Paul told me.

I looked around a little exasperated. It seems like the Massai hadn't seen or heard a lot of what I was talking about. "In my country, I can't go a day without hearing it. We hear it every day." I said. "Based on what I'm saying, would you ever want to move to America?"

They all completely broke down laughing again, this time even harder than before. They didn't have to think about the answer.

"They say 'no.' Never," Paul said.

The entire concept of transgenderism was so novel to them that the group of men I was talking to turned the script on me, and I become the subject of their interview. Obviously, the ideas I brought up were so unnatural that they had never crossed the minds of the Massai people. "How does it come from being a woman to a man again?" they asked.

"I don't know," I responded with all seriousness. "I guess they feel it. It's just something they feel. In my country a man will say, 'I'm a woman trapped in a man's body.'"

"How do you treat them?" someone in the back chimed in.

"Well," I responded, "Lots of people in my country say that if a man says he is a woman, we have to treat him like a woman."

"Does a man have breasts?" "Does he have a vagina?" "Does this man deliver?" "Does that man have a period?" The questions were rolling in, and I tried to answer as best I could. Soon, I was worried they were getting the wrong idea.

"Let me just clarify something. I don't have a woman trapped inside me. I'm just a regular man," I said as they all smiled, the laughter finally dying down.

Eventually the elder spoke to Paul. Paul turned back to me and said "What you are asking we have never seen, we have never heard. We believe if you are a man, you are a man. If you are a woman, you are a woman.... The truth that we believe is something that we have seen."

As I walked around the hot, dry, and dusty village where everyone was commonly swatting bugs away from their faces and where the sun felt like it was closer than anywhere else I'd ever been before, I couldn't help but notice that, nonetheless, everyone seemed happy. I thought about all the people in places like Europe and Canada and America who suffer from gender identity disorder or who have been taught to be confused about who they are and how they were born. I thought about the men and women who mutilate themselves in ever more novel and painful ways in vain attempts to make themselves happy. I thought about all the kids who are told over and over and over again that if they don't have their gender identity affirmed, then they will want to kill themselves. My heart was heavy at the giant weight of unhappiness that seemed to permeate so many lives back at home.

I decided to talk to the Massai women about this. According to my culture, if anybody had a reason to be unhappy, it would be them. They were not allowed to act at all like men, they were expected to bear children, and their duties were at home. Various people in the village had described women as a "helping hand to help a man" and as the neck of the body while men served as the head. That's not even to mention the complete lack of basic material comforts that everyone in the village, man and woman, experienced. I was sure they had some complaints.

"In my culture where I come from, lots of people are depressed. Do you have depression here?" I asked.

A woman named Mary responded. "No, no depression. People are very happy and friendly."

I looked at the poverty all around me. "Do you feel like you don't need to own a lot of things to be happy?" I asked.

"She said what makes them to be more happy," Paul began, "one, having kids.... Secondly, having cows, what you can provide, what cows produce, like milk, also they can make cream to give to the children. And also being in one place all sitting in a community, like a grove. That is the happiness too."

I must admit, I certainly wouldn't want to live like the Massai. I doubt any American who was honest with himself would. Their lives are terribly difficult. The poverty is visceral and always present. They certainly have some horrifying cultural practices that I do not condone. The truth is that after the Garden of Eden there has never been a place or a golden age when everything was perfect and made sense. But even so, the Massai people seem to have discovered a truth that we in the west don't know—or maybe they never forgot a truth that we no longer remember.

What is happiness? Well, think about what Mary said. Do you spend time with the ones you love? Do you have enough to provide for the basic needs of ones you love? Do you even have ones to love? Community. Basic sustenance. Family and children. He has most who needs least. He is happy who surrounds himself with the most love.

I had been asking the same question to everyone I had met for nearly a year, from doctors to therapists, from politicians to professors, from San Francisco to Kenya. The question "What is a woman?" is certainly about sex, gender, biology, social roles, and the like. Yet more profoundly, it is a question about identity. Where do we find our identity? How do we define ourselves? Is identity something we fulfill within the grand and ornate structure of nature, community, duties, and responsibilities—rooted in words like father, mother, son, daughter, friend, or child of God? Or is identity something we define for ourselves from within ourselves?

Maybe happiness comes not from making the world affirm "who we are," but by becoming who we were created to be.

NOTES

CHAPTER 1

[1] Planned Parenthood of Southeastern Pennsylvania v. Robert P. Casey, Legal Information Institute, Cornell Law School, June 29, 1992, https://www.law.cornell.edu/supct/html/91-744.ZO.html.

CHAPTER 2

[1] Gen. 5:2 (English Standard Version).

[2] Matt. 19:12 (English Standard Version).

[3] "Homily 62 on Matthew," New Advent, accessed January 14, 2022, https://www.newadvent.org/fathers/200162.htm.

[4] Deut. 23:1 (English Standard Version).

[5] Isa. 56:4-5 (English Standard Version).

[6] "Grace and Lace Letter: A Christian Publication for Crossdressers, Transgenderists, Transsexuals," Digital Transgender Archive, March 1994, https://www.digitaltransgenderarchive.net/files/2f75r825p.

[7] Ibid.

[8] Elizabeth Pérez, "'You Were Gonna Leave Them Out?': Locating Black Women in a Transfeminist Anthropology of Religion," *Feminist Anthropology* (2020), https://www.religion.ucsb.edu/wp-content/uploads/EPerez-You-Were-Gonna.pdf.

[9] Ibid.

[10] Jodi O'Brien, *Encyclopedia of Gender and Society* (Thousand Oaks: SAGE Publications, 2008), 64, https://books.google.com/books?id=_nyHS4WyUKEC&printsec=frontcover&source=gbs_ge_summary_r&cad=0#v=onepage&q=peoples%20view%20this%20as%20a%20Western%20concept&f=false.

[11] Matt Lebovic, "100 Years Ago, Germany's 'Einstein of Sex' Began the Gay Rights Movement," *Times of Israel*, November 11, 2019, https://www.timesofisrael.com/100-years-ago-germanys-einstein-of-sex-began-the-gay-rights-movement/.

[12] Magnus Hirschfeld, "Making Gay History," Making Gay History, podcast, accessed January 14, 2022, https://makinggayhistory.com/podcast/magnus-hirschfeld/.

[13] Joanne Meyerowitz, *How Sex Changed: A History of Transsexuality in the United States* (Cambridge: Harvard University Press, 2004), 19.

[14] Lebovic, "100 Years Ago, Germany's 'Einstein of Sex' Began the Gay Rights Movement."

[15] Ibid.

[16] Ibid.

17 Kevin Amidon, "Per Scientiam ad Justitiam: Magnus Hirschfeld's Episteme of Biological Publicity," Iowa State University, December 1, 2017, https://lib.dr .iastate.edu/language_pubs/139/.

18 Ralf Dose, "The World League for Sexual Reform: Some Possible Approaches," *Journal of the History of Sexuality* 12, no. 1 (2003): 1-15, https://www.jstor.org /stable/3704508.

19 "Weltliga Für Sexual Reform," on Internet Archive, accessed January 14, 2022, https://web.archive.org/web/20111122131239/http://www2.hu-berlin.de/sexology /gesund/archiv/gif/xwlsr_pl.jpg (the screenshot of the site was captured on November 22, 2011).

20 Meyerowitz, *How Sex Changed,* 15.

21 Ibid., 19.

22 Ibid.

23 Ibid., 26.

24 Benjamin Weinthal, "The Einstein of Sex at 150," Gay City News, May 10, 2018, https://www.gaycitynews.com/the-einstein-of-sex-at-150/.

25 Meyerowitz, *How Sex Changed,* 19.

26 Ibid., 15.

27 Ibid, 20.

28 Ibid., 19.

29 Weinthal, "The Einstein of Sex at 150."

30 Meyerowitz, *How Sex Changed,* 19.

31 Weinthal, "The Einstein of Sex at 150."

32 Meyerowitz, *How Sex Changed,* 45.

33 Ibid.

34 Ibid., 46.

35 Joanne Meyerowitz, "Sex Research at the Borders of Gender: Transvestites, Trans-sexuals, and Alfred C. Kinsey," *Bulletin of the History of Medicine* 75, no. 1 (2001), 83.

36 Meyerowitz, *How Sex Changed,* 46-47.

37 Mikella Procopio, "'Oh! Dr. Kinsey!': The Life and Work of America's Pioneer of Sexology," *Corinthian* 10, no. 9 (2009), https://kb.gcsu.edu/cgi/viewcontent.cgi? article=1078&context=thecorinthian.

38 Ibid., 293.

39 Ibid., 289.

40 Ibid.,292 & 297. Kinsey began at Indiana University in Bloomington after a ten month trip collecting wasps that began in 1919, meaning it his highly likely that he began his formal study of wasps in 1920. He ended his study of wasps in 1938.

41 Procopio, "'Oh! Dr. Kinsey!': The Life and Work of America's Pioneer of Sexology," 298.

42 Vern L. Bullough, "Alfred Kinsey and the Kinsey Report: Historical Overview and Lasting Contributions," *Journal of Sex Research* 35, no. 2, (1998), 129.

43 Procopio, "'Oh! Dr. Kinsey!': The Life and Work of America's Pioneer of Sexology," 299.

44 "The First Institute for Sexual Science (1919-1933)," Magnus-Hirschfeld-Gesellschaft e.V., accessed January 14, 2022, https://www.magnus-hirschfeld.de/ausstellungen /institute/.

45 Procopio, "'Oh! Dr. Kinsey!': The Life and Work of America's Pioneer of Sexology," 310.

46 Ibid., 300.

47 Ibid., 308.

48 Bullough, "Alfred Kinsey and the Kinsey Report," 130.

49 Procopio, "'Oh! Dr. Kinsey!': The Life and Work of America's Pioneer of Sexology," 308-309.

50 Bullough, "Alfred Kinsey and the Kinsey Report," 130.

51 Meyerowitz, *Sex Research at the Borders of Gender*, 85.

52 Ibid.

53 Ibid., 131.

54 Procopio, "'Oh! Dr. Kinsey!': The Life and Work of America's Pioneer of Sexology," 303.

55 Ibid., 300.

56 Caleb Crain, "Alfred Kinsey: Liberator or Pervert?" *New York Times*, October 3, 2004, https://www.nytimes.com/2004/10/03/movies/alfred-kinsey-liberator-or -pervert.html.

57 G.G. v. Gloucester County School Board, No. 15-2056 (2017), 10-11, https://www .aclu.org/sites/default/files/field_document/liberty_counsel_amicus.pdf.

58 Ibid.; Alfred C. Kinsey, Wardell B. Pomeroy, and Clyde E. Martin, *Sexual Behavior in the Human Male* (Bloomington: Indiana University Press, 1948), 160.

59 Ibid.

60 Kinsey, Pomeroy, and Martin, *Sexual Behavior in the Human Male*.

61 Bullough, "Alfred Kinsey and the Kinsey Report," 130.

62 Procopio, "'Oh! Dr. Kinsey!': The Life and Work of America's Pioneer of Sexology," 321.

63 Ibid., 321-322.

64 Richard Rhodes, "Father of the Sexual Revolution," *New York Times*, November 2, 1997, https://www.nytimes.com/1997/11/02/books/father-of-the-sexual-revolution .html.

65 Meyerowitz, *Sex Research at the Borders of Gender*, 88.

66 Ibid., 72.

67 Ibid., 74.

68 Ibid.

69 Ibid., 77.

70 Ibid., 87.

71 Verne L. Bullough, "The Contributions of John Money: A Personal View," *Journal of Sex Research* 40, no. 3 (2003), https://www.jstor.org/stable/3813317.

CHAPTER 3

1 Judith Reisman, *Kinsey: Crimes and Consequences: The Red Queen and the Grand Scheme* (Hartline Marketing, 1998), pdf p.170.

2 Ibid., 344.

3 "Learn Our History," Kinsey Institute, accessed January 14, 2022, https://kinsey institute.org/about/history/index.php.

4 "Program in Human Sexuality," University of Minnesota, on Internet Archive, accessed January 14, 2022, https://web.archive.org/web/20150724204551/http:/www.sexualhealth.umn.edu/education/john-money/bio (the screenshot of the site was captured on July 24, 2015).

5 Lisa Downing, Iain Morland, and Nikki Sullivan, *F*ckology: Critical Essays on John Money's Diagnostic Concepts* (Chicago: University of Chicago Press, 2014), 4.

6 Joanne Meyerowitz, *How Sex Changed*, 219.

7 Lawrence Lariar, *Oh! Dr. Kinsey!* Cartwrite Pub. Co. (1953) p. 295

8 Anke Ehrhardt, "John Money, Ph.D.", *The Journal of Sex Research* 44(3) (July 2007): 223-224 DOI:10.1080/00224490701580741

9 Downing, Morland, and Sullivan, *F*ckology*, 4.

10 John Colpatino, *As Nature Made Him* (New York: Harper Perennial, 2000), 27.

11 Benedict Carey, "John William Money, 84, Sexual Identity Researcher, Dies," *New York Times*, on archive.today, July 11, 2006, https://archive.md/sR2i6#selection-497.115-497.205 (the screenshot of the site was captured on January 4, 2013).

12 Ibid.

13 Ibid.

14 Ibid.

15 Ibid.

16 Lisa Downing, Iain Morland, and Nikki Sullivan, "Pervert or Sexual Libertarian?: Meet John Money, 'the Father of F***ology,'" *Salon*, January 4, 2015, https://www.salon.com/2015/01/04/pervert_or_sexual_libertarian_meet_john_money_the_father_of_fology/.

17 James Lincoln Collier, "Man and Woman Boy and Girl," *New York Times*, archives, February 25, 1973, https://www.nytimes.com/1973/02/25/archives/man-and-woman-boy-and-girl-by-john-money-and-anke-a-ehrhardt.html.

[18] J. Money, J.G. Hampson, and J.L. Hampson, "An Examination of Some Basic Sexual Concepts: The Evidence of Human Hermaphroditism," *Bull Johns Hopkins Hosp.* 97, no. 4 (1955), https://pubmed.ncbi.nlm.nih.gov/13260820/; Email message, "Money Partial Sourcing," November 22, 2021.

[19] Meyerowitz, *How Sex Changed*, 114.

[20] "Gender," Online Etymology Dictionary, accessed January 14, 2022, https://www.etymonline.com/word/gender.

[21] John Money, *Gay, Straight, and In-Between* (Oxford: Oxford University Press, 1988), 77, https://www.google.com/books/edition/Gay_Straight_and_In_between/8nTE_989I_cC?hl=en&gbpv=1&dq=%22Because+sex+differences+are+not+only+genitally+sexual,+although+they+may+be+secondarily+derived+from+the+procreative+organs,&pg=PA77&printsec=frontcover.

[22] Meyerowitz, *How Sex Changed*,114-115.

[23] Ibid., 125.

[24] Collier, "Man and Woman Boy and Girl."

[25] Ibid.

[26] Meyerowitz, *How Sex Changed*, 219.

[27] Downing, Morland, and Sullivan, "Pervert or Sexual Libertarian?"

[28] John Money, *Lovemaps: Clinical Concepts of Sexual/Erotic Health and Pathology, Paraphilia, and Gender Transposition in Childhood, Adolescence, and Maturity* (Buffalo: Prometheus Books, 2012), https://books.google.com/books/about/Lovemaps.html?id=2HsQ9FrHYZAC.

[29] John Money, *The Adam Principle* (Buffalo: Prometheus Books, 1993), 206.

[30] Ibid., 204.

[31] Ibid. , 206.

[32] Ibid., 204.

[33] "Famous Quotes on Boylove," ipce.info, accessed January 14, 2022, https://www.ipce.info/library/miscellaneous/famous-quotes-boylove.

[34] Reisman, *Kinsey: Crimes and Consequences*, 262-263.

[35] Meyerowitz, *How Sex Changed*, 218.

[36] Ibid., 219.

[37] Ibid.

[38] "Musing About Avon Wilson's Blended Life," TransGriot, April 5, 2009, https://transgriot.blogspot.com/2009/04/musing-about-avon-wilsons-blended-life.html.

[39] Downing, Morland, and Sullivan, *F*ckology*, 5.

[40] Ibid., 5.

[41] Meyerowitz, *How Sex Changed*, 142.

[42] Ibid., 220.

[43] Ibid.

44 Ibid., 222.

45 Jane E. Brody, "Benefits of Transsexual Surgery Disputed as Leading Hospital Halts the Procedure," *New York Times*, archives, October 2, 1979, https://www .nytimes.com/1979/10/02/archives/benefits-of-transsexual-surgery-disputed-as -leading-hospital-halts.html.

46 "Dr. Stanley Biber," History Colorado, June 26, 2020, https://www.historycolorado .org/story/2020/06/26/dr-stanley-biber.

47 Downing, Morland, and Sullivan, *F*ckology*, 72.

48 Ibid., 74.

49 Ibid.

50 Ibid.

51 Ibid., 78-79.

52 Ibid., 78.

53 Ibid.

54 Ibid.

55 Phil Gaetano, "David Reimer and John Money Gender Reassignment Controversy: The John/Joan Case," Embryo Project Encyclopedia, November 15, 2017, https:// embryo.asu.edu/pages/david-reimer-and-john-money-gender-reassignment -controversy-johnjoan-case.

56 Downing, Morland, and Sullivan, *F*ckology*, 70.

57 Ibid., 84.

58 Ibid., 85.

59 Ibid., 85-86.

60 Ibid., 87.

61 Ibid.

62 Ibid.

63 Gaetano, "David Reimer and John Money Gender Reassignment Controversy."

64 Downing, Morland, and Sullivan, *F*ckology*, 88.

65 Pam Rosenthal, "Forced Crossing," *Salon*, archives, February 24, 2000, https:// archive.md/YFe4q#selection-551.0-551.377.

66 Ibid.

67 Gaetano, "David Reimer and John Money Gender Reassignment Controversy."

68 "Our History," National Institute for the Study, Prevention, and Treatment of Sexual Trauma, accessed January 14, 2022, http://fredberlinmd.com/.

69 Gaetano, "David Reimer and John Money Gender Reassignment Controversy."

70 Downing, Morland, and Sullivan, F*ckology, 1.

71 Ehrhardt, Anke, "John Money, Ph.D.," *The Journal of Sex Research*, no. 3 (2007): 223-224, https://www.jstor.org/stable/20620298.

[72] Downing, Morland, and Sullivan, "Pervert or Sexual Libertarian?"

[73] Ibid.

[74] Ehrhardt, Anke, "John Money, Ph.D.," *The Journal of Sex Research,* no. 3 (2007): 223-224, https://www.jstor.org/stable/20620298.

CHAPTER 4

[1] "Doing Feminist Thinking with Judith Butler," Oxford Research Centre in the Humanities, May 28, 2021, https://www.torch.ox.ac.uk/event/the-politics-of-invisibility-a-conversation-with-judith-butler.

[2] Judith Butler, "Behavior Creates Your Gender," University of Oregon, Philosophical Installations, accessed January 14, 2022, https://philinstall.uoregon.edu/video/145/.

[3] "Judith Butler: Your Behavior Creates Your Gender: Big Think," YouTube video, posted by "Big Think," June 6, 2011, https://www.youtube.com/watch?v=Bo7o2LYATDc.

[4] "Gender Trouble," *New York Times,* on archive.today, March 12, 2006, https://archive.ph/x6wMi (the screenshot of the site was captured on February 24, 2021).

[5] Jocelyn Guest, "Mythbuster: The Scientist Who Exposed the Greatest Sexology Hoax of the 1970s is Back," University of Hawaii Manoa, Pacific Center for Sex and Society, accessed January 14, 2022, https://www.hawaii.edu/PCSS/biblio/articles/2005to2009/2006-mythbuster.html.

[6] Judith Butler, "Subjects of Sex/Gender/Desire," in *Feminisms* (New York: Oxford University Press, 1998), https://pressbooks.claremont.edu/clas112pomonavalentine/chapter/butler-judith-1998-subjects-of-sex-gender-desire/.

[7] Ani Ritchie and Meg Barker, "'There Aren't Words for What We Do or How We Feel So We Have to Make Them Up': Constructing Polyamorous Languages in a Culture of Compulsory Monogamy," *Sexualities* 9, no. 5 (2006), http://www.brown.uk.com/poly/ritchie.pdf.

[8] Email message from Justin Folk, "Gert's Office," November 18, 2021.

[9] "USA Today Says Women's and Gender Studies Is More Popular Than Ever!" University of Illinois Urbana-Champaign, Department of Gender & Women's Studies, March 5, 2017, https://gws.illinois.edu/news/2017-03-05/usa-today-says-womens-and-gender-studies-more-popular-ever.

[10] "18 Women's and Gender Studies Jobs," Indeed, April 29, 2021, https://www.indeed.com/career-advice/finding-a-job/womens-and-gender-studies-jobs.

[11] Cornog, Martha and Timothy Perper, *For SEX EDUCATION,* See *Librarian* (Greenwood Press, 1996), 30.

[12] "A Brief History of Sex Ed: How We Reached Today's Madness – Part II," Miriam Grossman M.D., June 6, 2013, https://www.miriamgrossmanmd.com/a-brief-history-of-sex-ed-how-we-reached-todays-madness-part-ii/.

[13] Reisman, *Kinsey: Crimes and Consequences,* 79.

[14] Ibid.

15 Joseph Epstein, "The Secret Life of Alfred Kinsey," *Commentary*, January 1998, https://www.commentary.org/articles/joseph-epstein/the-secret-life-of-alfred -kinsey/.

16 "Guidelines for Comprehensive Sexuality Education," Sexuality Information and Education Council of the United States, July 2018, https://siecus.org/wp-content /uploads/2018/07/Guidelines-CSE.pdf.

17 Ibid.

18 Robie H. Harris and Michael Emberley, *It's Perfectly Normal: Changing Bodies, Growing Up, Sex, and Sexual Health* (Somerville: Candlewick Press, 2014), https:// www.amazon.com/Its-Perfectly-Normal-Changing-Growing/dp/0763668729.

19 Cornog, Martha and Timothy Perper, *For SEX EDUCATION,* See *Librarian* (Greenwood Press, 1996), 179.

20 Charles Creitz, "Huckabee Blasts 'Embarrassing and Disgusting' NYC Private School's Explicit Sex-Ed for 1st Graders," Fox News, June 1, 2021, https://www .foxnews.com/media/huckabee-blasts-embarrassing-and-disgusting-nyc-private -schools-explicit-sex-ed-for-1st-graders.

21 Melissa Barnhart, "California's Sex Ed Guidelines Are 'Shocking' and 'Medically Risky' for Kids, Teacher Says," *Christian Post*, May 29, 2019, https://www.christianpost.com /news/californias-sex-ed-guidelines-shocking-medically-risky-for-kids-teacher-says .html.

22 Anna Anderson, "Sexually Explicit Books Were Put in These Virginia Classrooms. Parents Want Answers.," Daily Signal, November 2, 2019, https://www.daily signal.com/2019/11/02/sexually-explicit-books-were-put-in-these-virginia -classrooms-parents-want-answers/.

23 "Guidelines for Comprehensive Sexuality Education."

24 Cornog, Martha and Timothy Perper, *For SEX EDUCATION,* See *Librarian* (Greenwood Press, 1996), 29.

25 Ibid.

26 Ibid., 28.

27 Ibid., 40.

28 "Guidelines for Comprehensive Sexuality Education."

29 Anderson, "Sexually Explicit Books Were Put in These Virginia Classrooms. Parents Want Answers."

30 Katie Jerkovich, "Megyn Kelly Says Son's School Told Third Grade Boys About Taking Puberty Blockers So They Can Have Genitals 'Chopped Off,'" *Daily Caller*, October 20, 2021, https://dailycaller.com/2021/10/20/megyn-kelly-sons-school-told -boys-taking-puberty-blockers-can-have-genitals-chopped-off/.

31 Keri D. Ingraham, "The Radical Reshaping of K-12 Public Education: Gender Re- definition and Self-Selection," *American Spectator*, June 1, 2021, https://spectator .org/public-schools-gender-radical-reshaping/.

32 Ibid.

33 Homepage, Genderbread Person, accessed January 14, 2022, https://www
.genderbread.org/.

34 Ingraham, "The Radical Reshaping of K-12 Public Education: Gender Redefinition
and Self-Selection."

35 Sarah Taylor, "Parents Say Kindergarten and First-Grade Students Are Being Read
Wildly Age-Inappropriate Books About Sexuality and Gender Identity – and the
Schools Refuse to Let Them Opt Out," *Blaze Media*, November 16, 2021, https://www
.theblaze.com/news/parents-kindergarten-students-books-sexuality-gender-identity.

CHAPTER 5

1 Jan Hoffman, "Estimate of U.S. Transgender Population Doubles to 1.4 Million
Adults," *New York Times*, June 30, 2016, https://www.nytimes.com/2016/07/01
/health/transgender-population.html.

2 Meyerowitz, *How Sex Changed*, 1.

3 Ibid., 1.

4 Neil Amdur, "Renee Richards Ruled Eligible for U.S. Open," *New York Times*, on
archive.today, August 17, 1977, https://archive.ph/jCjkq (the screenshot of the site
was captured on September 30, 2019).

5 Associated Press, "A Look Back at the History of US Transgender Rights," Wesh 2
News, last updated February 23, 2017, https://www.wesh.com/article/a-look
-back-at-the-history-of-us-transgender-rights/8972867.

6 Jared Alexander, "Billy Porter on 'Cinderella' Role: I Dreamed of Being 'Male
Whitney Houston,'" Yahoo News, September 3, 2021, https://news.yahoo.com
/billy-porter-cinderella-role-dreamed-180800438.html.

7 Gwen Aviles, "Transgender Character Coming to Marvel Cinematic Universe,
Studio President Suggests," NBC News, last updated January 3, 2020, https://www
.nbcnews.com/feature/nbc-out/transgender-character-coming-marvel-cinematic
-universe-n1109521.

8 Matt Kane, "Glee Introduces Its First Transgender Character," Glaad (blog), April 18,
2012, https://www.glaad.org/blog/glee-introduces-its-first-transgender-character.

9 "Transparent," IMDb, accessed January 14, 2022, https://www.imdb.com/title
/tt3502262/?ref_%3Dnv_sr_1.

10 "6 TV Shows with Transgender Characters Played by Transgender Actors," Hypable,
June 6, 2019, https://www.hypable.com/tv-shows-transgender-characters-played
-by-transgender-actors/.

11 Paulina Firozi, "Singer Demi Lovato Says They Identify As Nonbinary: 'This Is My
Truth and I Can't Shove It Down," *Washington Post*, May 19, 2021, https://www.wash
ingtonpost.com/arts-entertainment/2021/05/19/demi-lovato-nonbinary-identity/.

12 Cassandra Fairbanks, "Former Disney Star Demi Lovato Tells Followers to 'Be
a Sl*t' and 'Make Porn,'" Gateway Pundit, September 8, 2021, https://www.the
gatewaypundit.com/2021/09/former-disney-star-demi-lovato-tells-followers
-slt-make-porn/.

13 Calvin Freiburger, "'Good Morning America' Promotes Child Drag Queen," *LifeSiteNews*, November 16, 2018, https://www.lifesitenews.com/news/good-morning-america-promotes-child-drag-queen/.

14 Ibid.

15 Gwen Aviles, "Trans Woman Is First Rhodes Scholar in Program's 117-Year History," NBC News, November 25, 2019, https://www.nbcnews.com/feature/nbc-out/trans-woman-first-rhodes-scholar-program-s-117-year-history-n1090866.

16 Sam Levine, "9-Year-Old Becomes First Openly Transgender Boy Scout," *Huffington Post*, February 8, 2017, https://www.huffpost.com/entry/transgender-boy-scouts_n_589b3bd0e4b04061313a959d.

17 Katy Steinmetz, "Meet the First Openly Transgender Speaker at a Party Convention," *Time*, July 26, 2016, https://time.com/4422613/democratic-convention-first-transgender-sarah-mcbride/.

18 Julie Compton, "'Boy or Girl?' Parents Raising 'Theybies' Let Kids Decide," NBC News, July 19, 2018, https://www.nbcnews.com/feature/nbc-out/boy-or-girl-parents-raising-theybies-let-kids-decide-n891836.

19 Abigail Shrier, "When Your Daughter Defies Biology," *Wall Street Journal*, January 6, 2019, https://www.wsj.com/articles/when-your-daughter-defies-biology-11546804848.

20 Lisa Littman, "Parent Reports of Adolescents and Young Adults Perceived to Show Signs of a Rapid Onset of Gender Dysphoria," *Plos One* (2018), https://journals.plos.org/plosone/article?id=10.1371/journal.pone.0202330.

21 Ibid.

22 Michelle Cretella, "I'm a Pediatrician. How Transgender Ideology Has Infiltrated My Field and Produced Large-Scale Child Abuse.," *Daily Signal*, July 3, 2017, https://www.dailysignal.com/2017/07/03/im-pediatrician-transgender-ideology-infiltrated-field-produced-large-scale-child-abuse/.

CHAPTER 6

1 "A Guide to Hormone Therapy for Trans People," COI, 2007, https://www.scottishtrans.org/wp-content/uploads/2013/06/NHS-A-Guide-to-Hormone-Therapy-for-Trans-People.pdf.

2 Ibid.

3 M. Klaver et al., "Cross-Sex Hormone Therapy in Transgender Persons Affects Total Body Weight, Body Fat and Lean Body Mass: A Meta-Analysis," *Andrologia* 49, no. 5 (2016), https://pubmed.ncbi.nlm.nih.gov/27572683/.

4 "A Guide to Hormone Therapy for Trans People," 10.

5 "What to Expect: Vaginoplasty at the University of Michigan Health System," University of Michigan Health System, last updated December 2016, https://www.med.umich.edu/pdf/Vaginoplasty.pdf.

6 Ibid.

7 Ibid.

[8] de Graaf, Natasja M., et al, Suicidality in clinic-referred transgender adolescents," *European Child & Adolescent Psychiatry* (2020), https://doi.org/10.1007/s00787 -020-01663-9.

CHAPTER 7

[1] American Heart Association, "Hormone therapy may increase cardiovascular risk during gender transition." ScienceDaily, February 18, 2019. www.sciencedaily .com/releases/2019/02/190218093959.htm.

[2] Scott Newgent, "Forget What Gender Activists Tell You. Here's What Medical Transition Looks Like," *Quillette,* October 6, 2020, https://quillette.com/2020/10/06/ forget-what-gender-activists-tell-you-heres-what-medical-transition-looks-like/.

[3] "Transgender Healthcare," Planned Parenthood, Accessed January 7, 2022, https:// www.plannedparenthood.org/planned-parenthood-greater-texas/patient -resources/transgender-healthcare.

[4] Susan Berry, "FDA: Thousands of Deaths Linked to Puberty Blockers," *Breitbart,* October 2, 2019, https://www.breitbart.com/politics/2019/10/02/fda-thousands-of -deaths-linked-to-puberty-blockers/

[5] Christina Jewett, "Women Fear Drug They Used to Halt Puberty Led to Health Problems," *Kaiser Health. News,* February 2, 2017, https://khn.org/news/women -fear-drug-they-used-to-halt-puberty-led-to-health-problems/.

[6] Joe Cochrane, "Indonesia Approves Castration for Sex Offenders Who Prey on Children," *New York Times,* May 25, 2016, https://www.nytimes.com/2016/05/26 /world/asia/indonesia-chemical-castration.html.

[7] Carmichael, et al. "Short-term outcomes of pubertal suppression in a selected cohort of 12 to 15 year old young people with persistent gender dysphoria in the UK," *Plos One* 16(2): e0243894 (2021), https://journals.plos.org/plosone /article?id=10.1371/journal.pone.0243894.

[8] L.C. de Vries, et al. "Puberty Suppression in Adolescents with Gender Identity Disorder: A Prospective Follow-Up Study," *The Journal of Sexual Medicine* (2011), https://www.jsm.jsexmed.org/article/S1743-6095(15)33617-1/fulltext.

[9] Michael Biggs, "Puberty Blockers and Suicidality in Adolescents Suffering from Gender Dysphoria." *Archives of Sexual Behavior* 227–2229 (2020), https://doi .org/10.1007/s10508-020-01743-6.

[10] Turban, et al., "Pubertal Suppression for Transgender Youth and Risk of Suicidal Ideation," *Pediatrics* (2020), https://publications.aap.org/pediatrics/article/145/2/ e20191725/68259/Pubertal-Suppression-for-Transgender-Youth-and.

[11] Ibid., Table 3.

[12] Carmichael, et al., "Short-term outcomes of pubertal suppression in a selected cohort of 12 to 15 year old young people with persistent gender dysphoria in the UK."

[13] Brandon Showalter, "NHS removes trans guidance claim that puberty blockers are 'fully reversible,'" *Christian Post,* June 6, 2020, https://www.christianpost.com /news/nhs-removes-trans-guidance-claim-that-puberty-blockers-are-fully -reversible-adds-list-of-risks.html.

[14] Riittakerttu, et al., "Adolescent development and psychosocial functioning after starting cross-sex hormones for gender dysphoria." *Nordic Journal of Psychiatry* (2020), https://www.tandfonline.com/action/showCitFormats?doi=10.1080%2F08039488.2019.1691260.

[15] "Medical treatment methods for dysphoria associated with variations in gender identity in minors – recommendations," Palveluvalikoima, January 16, 2020, https://palveluvalikoima.fi/documents/1237350/22895008/Summary_minors_en.pdf/aaf9a6e7-b970-9de9-165c-abedfae46f2e/Summary_minors_en.pdf.

[16] Lia Nainggolan, "Hormonal TX of Youth with Gender Dysphoria Stops in Sweden," *Medscape*, May 12, 2021, https://www.medscape.com/viewarticle/950964.

[17] Lisa Nainggolan, "AAP 'Silencing Debate' on Gender Dysphoria, Say Doctor Group," *Medscape*, August 16, 2021, https://www.medscape.com/viewarticle/956650.

[18] Carmichael, et al., "Short-term outcomes of pubertal suppression in a selected cohort of 12 to 15 year old young people with persistent gender dysphoria in the UK."

[19] Scott Newgent, "Forget What Gender Activists Tell You. Here's What Medical Transition Looks Like,"

[20] Rachel Savage and Hugo Greenhalgh, "UK court rules against trans clinic over treatment for children," Reuters, November 30, 2020, https://www.reuters.com/article/us-britain-lgbt-transgender-trfn-idUSKBN28B3AV.

[21] Haroon Siddique, "Appeal court overturns UK puberty blockers ruling for under-16s," *The Guardian*, September 17, 2021, https://www.theguardian.com/society/2021/sep/17/appeal-court-overturns-uk-puberty-blockers-ruling-for-under-16s-tavistock-keira-bell.

[22] Rachid, Mamoon and Muhammad Sarmad Tamimy, "Phalloplasty: The dream and the reality," *Indian Journal of Plastic Surgery* May-Aug; 46(2): 283–293 (2013), https://www.ncbi.nlm.nih.gov/pmc/articles/PMC3901910/.

[23] "Transgender health care: Does health insurance cover gender-affirming surgery?" *HealthSherpa*, Accessed January 17, 2021, https://blog.healthsherpa.com/transgender-health-care-gender-affirming-surgery-health-insurance.

[24] Madeleine Kearns, "The grim reality of gender reassignment," *Spectator Australia*, November 14, 2021. https://www.spectator.com.au/2021/11/the-grim-reality-of-gender-reassignment/.

[25] "Why the Growth of Transgender Surgery Centers in the U.S. Matters," *Trans Health Care*, November 11, 2019, https://www.transhealthcare.org/news/transgender-surgery-centers/.

CHAPTER 8

[1] Warner Todd Huston, "'So Easy I was Cruising': Trans U. Penn Swimmer Bragged About Beating Female Opponents," *Breitbart*, December 13, 2021, https://www.breitbart.com/sports/2021/12/13/so-easy-i-was-cruising-trans-upenn-swimmer-bragged-about-beating-female-opponents/.

[2] Ryan Grenoble, "ESPN's Curt Schilling Goes on Anti-Transgender Rant," *Huffpost*, April 19, 2016, https://www.huffpost.com/entry/curt-schilling-transgender-rant _n_571690bee4b0018f9cbb713b.

[3] Cover Media, "JK Rowling now 'she who must not be named' for Harry Potter fans," *Newshub*. April 7, 2020, https://www.newshub.co.nz/home/entertainment/2020/07 /jk-rowling-now-she-who-must-not-be-named-for-harry-potter-fans.html.

[4] Joran Moreau, "J.K. Rowling Gets Backlash Over Anti-Trans Tweets," *Variety*, June 6, 2020, https://variety.com/2020/film/news/jk-rowling-transphobic-tweets -controversy-1234627081/#!.

[5] Abigail Shrier, "Gender activists are trying to cancel my book; Why is Silicon Valley helping them?" *Pittsburgh Post-Gazette*, November 24, 2020, https://www .post-gazette.com/opinion/2020/11/22/Gender-activists-Silicon-Valley-Transgender -LGBTQ/stories/202011220021.

[6] Ibid.

[7] Ibid.

[8] Matthew Wilson, "Conservative journalist's Princeton talk given in secret location as students protest, denounce event," *The College Fix*, December 10, 2021, https:// www.thecollegefix.com/conservative-journalists-princeton-talk-given-in-secret -location-as-students-protest-denounce-event/.

[9] Brandon Showalter, "GoFundMe takes down 'puberty is not a medical condition' billboard campaign; new effort launched," *Christian Post*, on Internet Archive, accessed February 10, 2022, https://archive.fo/PUWZo (the screenshot of the site was captured on February 10, 2022).

[10] Lisa Littman, "Correction: Parent reports of adolescents and young adults perceived to show signs of a rapid onset of gender dysphoria," *Plos One* (2019), https://www.ncbi.nlm.nih.gov/pmc/articles/PMC6424391/.

[11] "Scott Newgent on his permanent ban from Twitter," YouTube video, posted by "The Center for Bioethics and Culture Network," August 8, 2021, https://www .youtube.com/watch?v=ziBc6HYHVUM.

[12] Tiramillas, "Netflix proud as Ellen Page announces she is transgender and changes name to Elliot," *Marca*, February 12, 2020, https://www.marca.com/en/lifestyle /2020/12/02/5fc80e4e268e3ec5518b468a.html.

[13] Laurel Wamsley, "Pentagon Releases New Policies Enabling Transgender People to Serve In The Military," NPR, March 31, 2021, https://www.npr.org/2021/03 /31/983118029/pentagon-releases-new-policies-enabling-transgender-people-to -serve-in-the-milit.

[14] Abigail Shrier, "How Activist Teachers Recruit Kids," *Substack*, November 18, 2021, https://abigailshrier.substack.com/p/how-activist-teachers-recruit-kids.

[15] Sarah Parshall Perry and Alexander Phipps, "School Districts Are Hiding Information About Gender-Transitioning Children From Their Parents. This Is Unconstitutional." *The Daily Signal*, March 24, 2021, https://www.dailysignal.com/2021/03/24 /school-districts-are-hiding-information-about-gender-transitioning-children-from -their-parents-this-is-unconstitutional/.

16 Steve Stewart, "Leon School Officials Develop Gender Transition Plan Without Parent Approval," *Tallahassee Reports*, September 26, 2021, https://tallahasseereports.com/2021/09/26/leon-school-officials-develop-gender-transition-plan-without-parent-approval/.

17 Perry and Phipps, "School Districts Are Hiding Information About Gender-Transitioning Children From Their Parents. This Is Unconstitutional."

18 Tyler Arnold, "Virginia teacher quits after Loudon County schools adopts transgender policy," on Internet Archive, accessed February 10, 2022, https://web.archive.org/web/20210817133752/https://www.insidenova.com/news/state/virginia-teacher-quits-after-loudoun-county-schools-adopts-transgender-policy/article_3727b9b6-26cf-5fdc-9bfc-c3336b07381c.html (the screenshot of the site was captured on August 17, 2021).

19 John Riley, "Court rules against Indiana teacher who resigned rather than call trans students by their chosen names," on Internet Archive, accessed February 10, 2022, https://web.archive.org/web/20210717045022/https://www.metroweekly.com/2021/07/court-rules-against-indiana-teacher-who-resigned-rather-than-call-trans-students-by-their-chosen-names/ (the screenshot of the site was captured on July 17, 2021).

20 Crystal Mundy, "Pedophilia as Age Sexual Orientation: Supporting Seto's (2012) Conceptualization," *ResearchGate* (2020), https://www.researchgate.net/publication/345918300_Pedophilia_as_Age_Sexual_Orientation_Supporting_Seto%27s_2012_Conceptualization.

21 "ODU professor: The word 'pedophile' shouldn't be used to refer to people attracted to children," YouTube video, posted by "13News Now, November 15, 2021, https://www.youtube.com/watch?v=UNW13jjSL-E.

22 Andy Ngo, "Sex offending suspect claims transgender harassment in Wi Spa Case," *New York Post*, September 2, 2021, https://nypost.com/2021/09/02/charges-filed-against-sex-offender-in-wi-spa-casecharges-filed-against-sex-offender-in-notorious-wi-spa-incident/.

CHAPTER 9

1 Bruce Bawer, "'A Certain Madness Amok,'" *City Journal*, April 1, 2021, https://www.city-journal.org/canadian-father-jailed-for-speaking-out-about-trans-identifying-child.

2 "ACLU-WA and Peacehealth Agree to Settle Lawsuit Involving Transgender Healthcare," ACLU, January 3, 2019, https://www.aclu.org/press-releases/aclu-wa-and-peacehealth-agree-settle-lawsuit-involving-transgender-healthcare.

3 David Lee, "Father Who Lost Custody of Trans Child Runs for Texas House to Outlaw Child Gender Reassignments," Courthouse News Service, December 8, 2021, https://www.courthousenews.com/father-who-lost-custody-of-trans-child-runs-for-texas-house-to-outlaw-child-gender-reassignments/.

4 Jen Christensen, "Judge Gives Grandparents Custody of Ohio Transgender Teen," CNN Health, last updated February 16, 2018, https://www.cnn.com/2018/02/16/health/ohio-transgender-teen-hearing-judge-decision/index.html.

5 Rod Dreher, "ROGD Hell," *American Conservative*, January 7, 2019, https://www.theamericanconservative.com/dreher/rapid-onset-gender-dysphoria-hell/.

6 Abigail Shrier, "When the State Comes for Your Kids," *City Journal*, June 8, 2021, https://www.city-journal.org/transgender-identifying-adolescents-threats-to-parental-rights.

7 Ibid.

CHAPTER 10

1 Charlie Jacobs, "What I've Learned Rescuing My Daughter from Her Transgender Fantasy," *Daily Signal*, December 13, 2021, https://www.dailysignal.com/2021/12/13/what-ive-learned-rescuing-my-daughter-from-her-transgender-fantasy/.

2 Laurel Duggan, "'Predators' and 'Child Abusers': Matt Walsh Berates the Loudoun County School Board," *Daily Caller*, September 29, 2021, https://dailycaller.com/2021/09/29/speech-matt-walsh-loudoun-county-virginia-school-board/.

3 Mary Margaret Olohan and Kendall Tietz, "EXCLUSIVE: Crowds Gather with Matt Walsh to Protest Virginia School's 'Indoctrination and Psychological Abuse of Kids,'" *Daily Caller*, September 28, 2021, https://dailycaller.com/2021/09/28/matt-walsh-loudoun-county-virginia-school-critical-race-theory-transgender/.